AF698208

WITHDRAWN

ADVANCES IN THE ECONOMICS OF ENERGY AND RESOURCES

Volume 2 • 1979

THE PRODUCTION AND PRICING OF ENERGY RESOURCES

ADVANCES IN THE ECONOMICS OF ENERGY AND RESOURCES

A Research Annual

Editor: ROBERT S. PINDYCK
Sloan School of Management
Massachusetts Institute of Technology

VOLUME 2 • 1979

THE PRODUCTION AND PRICING OF ENERGY RESOURCES

JAI PRESS INC.
Greenwich, Connecticut

Copyright © 1979 JAI PRESS INC.
165 West Putnam Avenue
Greenwich, Connecticut 06830
All rights reserved. No part of this publication may be reproduced, stored on a retrieval system, or transmitted in any form or by any means, electronic, mechanical, photocopying, filming, recording or otherwise without prior permission in writing from the publisher.

ISBN NUMBER: 0-89232-079-6

Manufactured in the United States of America

HD
9502
.A1
A3
v.2

CONTENTS

INTRODUCTION

Robert S. Pindyck

Advances in the Economics of Energy and Resources brings together new research results covering a wide spectrum of problems in energy and resource economics and policy. In this second volume, *The Production and Pricing of Energy Resources*, ten papers have been assembled that cover a wide range of problems in energy resource supply, pricing, and management. Topics treated in this volume include the statistical forecasting of resource discovery rates, the optimal exploitation and pricing of resources, the role of supply in world resource markets, and the relationship between resource markets and the macroeconomy. In editing *Advances in the Economics of Energy and Resources*, this breadth of coverage was one objective, but the more important objective was obtaining a collection of papers of uniformly high quality. I believe that this objective was met, and that the papers here represent new and significant research results in the economics of energy and resources.

Forecasting oil supplies is of crucial concern to both the energy industry, engaged in exploration, and the government, engaged in formulating energy policy. Furthermore, forecasts of reserves and production have been the focus of considerable debate; witness the recent arguments over conflicting estimates of "potential" oil and gas reserves in the United States and around the world. The first paper in this volume, "Alternative Methods of Oil Supply Forecasting," by M. A. Adelman and Henry Jacoby, describes the theory and application of two alternative approaches to oil supply forecasting. Specifically, Adelman and Jacoby are concerned with predicting the new reserves of oil and the production of oil out of reserves that are likely to result from exploratory and development activity. The models they construct are intended to predict oil supplies in different regions of the world for non-OPEC "price-taker" countries (and ultimately to be part of a larger simulation model of the world oil market). After surveying previous approaches to energy supply forecasting, Adelman and Jacoby describe two approaches that combine geological and economic information. The first approach, "Disaggregated Pool Analysis," is the most detailed (and requires the most data), and is based on a statistical analysis of the exploratory process, and an economic analysis of the pools that are predicted to be found. In this approach, forecasts are generated not only of the total recoverable reserves to be discovered by a given level of exploratory activity, but also of the distribution of pool size and the sequence of discoveries by size. The second approach to be used when the detailed data needed for the disaggregated analysis is not available, is based on forecasts of rig activity and the analysis of proved reserves added per rig year. The forecasted reserve additions in turn become an input to forecasts of capacity expansion and likely oil production. Adelman and Jacoby apply their "Disaggregated Pool Analysis" to recent oil discoveries in the North Sea, and their aggregated approach to a number of oil-producing countries.

The paper by Paul Eckbo is also concerned with forecasting the response of reserve discoveries to exploration. His paper, "A Basin Development Model of Oil Supply," presents a model that can be used to estimate the supply potential of a region given an exogenously specified time profile for exploratory drilling. The model, which operates at the basin level (a basin is defined as "a continuous containment of possible hydrocarbon source rock," and will usually contain a number of plays), relies on a deterministic discovery decline relationship to generate an expected discovery sequence, which in turn serves as a first approximation to the joint analysis of the exploration foreplays and reservoirs inside a play. In Eckbo's model, the reservoirs found enter into a reservoir submodel, which takes account of costs and expected future prices, and allows detailed consideration of the tax regime. By separating exploration and finding activities from development and production activities Eckbo's model allows consideration of the two major aspects of resource depletion—the depletion of producible reservoirs from the population of reservoirs to be found, and the depletion of recoverable reserves from the existing population of producible reservoirs. Eckbo applies his model to data for oil discoveries in the North Sea, and shows that it provides a quite useful forecasting tool.

The paper by Martin Zimmerman, "A Policy Model of U.S. Coal Supply," is also concerned with forecasting resource supply. Zimmerman's focus is the long-run supply of coal, and he estimates a "cumulative cost function" that determines the long-run marginal cost of mining as a function of cumulative output produced over time. In his procedure, he relates cost to a rate of output and geological conditions, and then combines geological information on reserves with the cost function to yield the cumulative cost function. Zimmerman's paper begins with a discussion of coal mining technology that provides a basis for the cost estimation. He next presents estimations of cost functions for both strip-mining and for deep-mining, and then combines the available data on remaining coal deposits to yield cumulative cost functions. Zimmerman concludes with a discussion of the likely effects of environmental regulation on the evolution of coal costs.

The next five papers in this volume all deal with theoretical issues in the exploration, production, and pricing of exhaustible resources. The paper by Russell Uhler, "The Rate of Petroleum Exploration and Extraction," attempts first to study the optimal path of exploration in a petroleum play in which geological information is accumulated and aids in reservoir discovery, but where discoveries also exhaust the stock of undiscovered reserves. After obtaining a solution to this problem, Uhler uses parameter estimates to generate the optimal path of exploration in one region in Alberta, Canada, and then compares this path with actual industry behavior. Next, Uhler examines the problem of extraction from known reservoirs. The problem of optimal extraction of an exhaustible resource is certainly not a new one, but Uhler's approach is unique in that he integrates the physical behavior of reservoirs into the economic model of extraction, thereby combining an economic and engineering approach to the problem. Uhler shows that the optimal extraction rate is highly dependent on the pressure conditions within the reservoir, and he goes on to consider the effects of an investment in pressure maintenance facilities.

In "Uncertainty and the Optimal Supply Policy for an Exhaustible Resource," Geoffrey Heal determines the effects of uncertainty on the socially optimal rate of exploitation of an exhaustible resource. In particular, he considers three alternative cases in which uncertainty can arise. (1) existing reserves of a resource are known with certainty, but additional reserves will be discovered in the future with uncertain timing and magnitude; (2) existing reserves of a resource are known with certainty, but a substitute for the resource will be developed at an uncertain time in the future and with an uncertain scale of availability; and (3) there is uncertainty about the size of existing reserves of the resource. For each of these cases, Heal first demonstrates the characteristics of the solution when there is no uncertainty (e.g., the size and timing of future reserve discoveries are known exactly), and then shows how the solution changes when varying degrees of uncertainty are introduced. Heal shows that under some conditions, the effect of uncertainty can be approximated (and for policy purposes treated) using a certainty-equivalent result of an increase in the discount rate. And he shows that in some cases the effects of uncertainty are quite counter-intuitive; in case (3), for example, a mean-preserving increase in the uncertainty about the initial reserves of the resource leads to an *increase*, rather than decrease, in the initial level of resource use.

Richard Gilbert also examines the effects of uncertainty in his paper, "Search Strategies and Private Incentives for Resource Exploration." Gilbert's objective, however, is to determine how, given uncertainty, exploratory activity would be used by private producers to increase their reserve levels (thus allowing for increased future production), and to reduce the existing uncertainty (about the size of the potential stock, the cost of extraction, and the location of deposits). Beginning with a simplified model of exploration in which extraction costs are constant and discovery probability distributions do not change as exploratory activity goes on, Gilbert shows that exploration is equivalent to extraction, and unexplored areas can be characterized by a certainty-equivalent extraction cost, so that it is optimal to simply exploit areas in order of increasing costs. He then goes on to analyze the speculative value of information resulting from exploration, and examines the effects of speculation (as in competitive lease auctions) on the private incentives for exploratory activity.

Donald Hanson, in "Increasing Extraction Costs and Resource Prices," provides a fresh look at the role of extraction costs on the optimal production and price paths of an exhaustible resource. Hanson examines resource use in an economic growth context, and shows (for a model assuming perfect certainty) that even when extraction costs are increasing, the efficiency price of the extracted resource currently in use satisfies the same relationship as in the conventional theory where extraction cost is constant. In effect the growth rate of the efficiency price equals the interest rate times the fraction of the efficiency price that is not extraction cost. Hanson also examines the degree to which the efficiency price of the extracted resource follows the extraction cost path. Finally, Hanson shows that the efficiency price path is also a competitive equilibrium with perfect foresight.

One reason for OPEC's recent success in vastly increasing its oil revenues is that no "backstop" technology exists for producing an alternative to oil at a cost close to today's price of oil. However, should OPEC raise the price of oil further (or should a new backstop technology by introduced that will produce an economical alternative to oil), the existence of a substitute will make the demand for OPEC oil much more sensitive to price changes than has been the case in the past. In "Staving Off the Backstop: Dynamic Limit-Pricing with a Kinked Demand Curve," Stephen Salant examines OPEC's optimal pricing policy over time given the existence of a high-cost backstop technology. Assuming that OPEC's goal is to maximize its discounted stream of profits, Salant considers whether it is in OPEC's interest to sell oil only at prices below the limit price at which the backstop becomes economical, or to sell for some period of time at the limit price. As Salant points out, the solution to the cartel's pricing problem is important not only to the cartel members, but also to the importing countries, since it enables them to forecast the cartel price more accurately, and to react more rationally.

The next paper, "ETA-MACRO: A Model of Energy/Economy Interactions," is by Alan Manne, and describes the functioning and application of his model of the energy sector and its relationship to long-run economic growth. ETA-MACRO is an integrated model that operates in a dynamic nonlinear optimization framework. It incorporates the impending exhaustion of oil and gas reserves (and a transition to alternative energy supplies over the next twenty to fifty years), the effects of price on conservation, and the effects of rising energy costs on the accumulation of physical capital in future time periods. The model contains two dynamic submodels, a process analysis model for technology assessment, and a macroeconomic growth model that allows for substitution between capital, labor and energy inputs. After describing the model's structure, Manne applies the model to the analysis of a particular energy policy option—a moratorium on the introduction of any new civilian nuclear power plants in the United States. Manne shows that, according to the model, a "no nuclear" policy would have negligible macroeconomic effects prior to the year 2000, but would lead to an annual loss of $100 billion (about 3 percent of the GNP) by the year 2000.

In the last paper of the volume, Harold Bernett examines the impact of the recent cartelization of the world market of an important nonenergy resource: bauxite. In "The Bauxite Cartel in the New International Economic Order," Bernett discusses the meaning of the recent "New Economic Order" proposals of the Third World countries and the United Nations, and argues that the International Bauxite Association (IBA) cartel was formed, based on New Economic Order concepts, to emulate the OPEC cartel. Bernett explains, however, why IBA is unlikely to be as successful as OPEC over a significant period of time. Barnett argues that relative to bauxite needs, there are too many alternative sources of supply of bauxite, and also of alternative aluminum-bearing ores, to allow the cartel to continue to control supply and price. Barnett concludes by suggesting some scenarios by which the cartel might weaken or break down.

ALTERNATIVE METHODS OF OIL SUPPLY FORECASTING

M. A. Adelman, MASSACHUSETTS INSTITUTE OF TECHNOLOGY

H. D. Jacoby, MASSACHUSETTS INSTITUTE OF TECHNOLOGY

I. THE ROLE OF "PRICE-TAKER" SUPPLIERS IN THE OIL MARKET

Analysis of likely developments in the world oil market is ultimately dependent on some method of forecasting oil supply from key regions. Unfortunately, data problems tend to dominate work in this area, and much of the analysis task reduces to making the best use of the limited information that is available. Here we report on two alternative approaches to this forecasting problem, both avowedly data-oriented.

Petroleum exporters need to be grouped into two rough categories. First, there are what we will call "price-taker" suppliers. This is a group of

Advances in the Economics of Energy and Resources, Vol. 2, pp. 1–38.
Copyright © 1979 by JAI Press, Inc.
All rights of reproduction in any form reserved.
ISBN: 0-89232-079-6

petroleum exporters who appear to act as *price-takers* in the sense that each takes the world price (which is being set by others) as given. Each makes supply decisions according to his own parochial interest, without concern for their impact on the world price. This group includes non-OPEC sources of the North Sea, the USSR, China, and Mexico. It also may include members of OPEC who have low per capita incomes such as Algeria, Indonesia, and Nigeria. Second, there is the "cartel core"—a small group of nations who are the *price-makers*. This core includes Saudi Arabia, Kuwait, and others on the Arabian Peninsula; it also may include Iraq, Iran, Libya, and Venezuela. These countries face a residual demand for world oil, which is the total demand less that supplied by the price-taker exporters.

These groupings are not hard and fast; indeed a major focus of our inquiry is the circumstance in which a given exporter would change from one to another camp. The world oil scene is a dynamic interplay among these importers and suppliers wherein the oil price is set by the members of the cartel core, who assume the task of controlling oil production so that it does not outstrip the world demand forthcoming at that price.

In this paper our focus is on the price-takers. And, since the desired form of a supply function depends on its intended use, we begin with a brief look at the broader market studies for which these supply analysis methods are designed. The structure of the overall study is shown in Figure 1; the figure also is a simple flow diagram of the simulation model framework we are using to tie the various pieces of work together.[1] The heart of the project is the supply and demand studies shown in the middle of the figure. These studies seek to improve our understanding of the fundamental market forces, and to provide estimates of supply functions for price-taker suppliers and demand functions for importers. These functions are then incorporated into a simulation model of overall market performance.

The simulation framework is designed to accept an anticipated oil price trajectory, and to compute the resulting demands, supplies, and other market characteristics over the study period. Hypotheses about likely cartel price behavior are developed using a separate set of behavioral models,[2] as shown in the cloud at the upper left of Figure 1.

Thus we approach the problem with two types of models—analytical representations of cartel behavior, and a detailed simulation of market supply and demand. The reason for the division is analytical convenience. The determinants of import demand and price-taker supply are varied and complex; they involve cost and price, along with the effects of tax and regulatory policies. To analyze the likely response of the market to one or another price pattern, one needs a method that can accept unwieldy functional relationships. This requirement leads to a simulation framework for the overall analysis of market demand and supply outside the cartel.

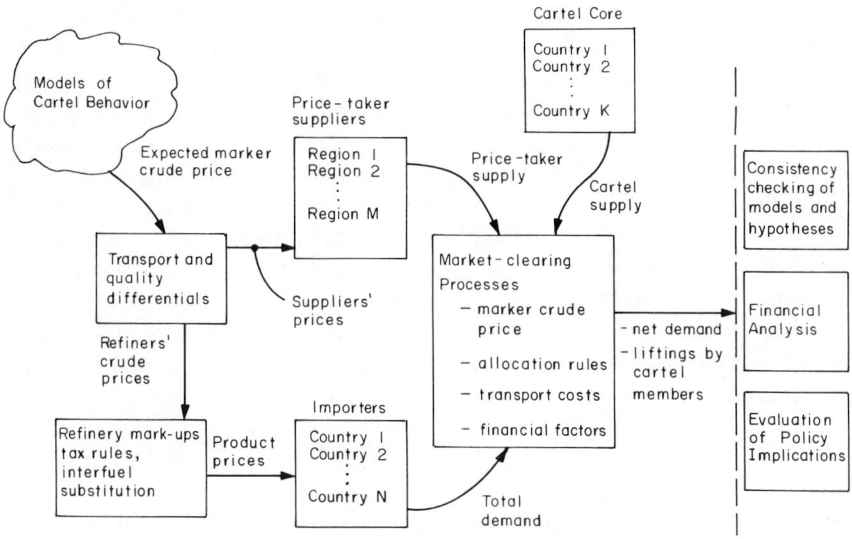

Figure 1. Overall analysis framework.

On the other hand, study of the cartel itself, and its pricing decisions, often involves some form of static or dynamic optimization calculation. For this part of the analysis, drastically simplified supply-demand relationships are needed so that many formulations of cartel behavior may be simply and cheaply tested. The two analyses feed one another, as shown in Figure 1.

In keeping with our emphasis on the underlying forces in the market, the simulation framework is based on what we call a "bathtub" approximation to the world oil market. That is, the market is treated as a single pool, where exporters put oil in and importers draw it out. The details of the transportation network and the refinery and distribution sector are almost neglected. Our aim is to match demand for *products* with the supply of *crude*, treating the intervening margin as a buffer, exogenously determined. Data and simple models of these factors are part of the simulation framework, as shown in the two boxes at the left of Figure 1. We plan to add more complex representations of these subsectors only as necessary.

The result of the overall simulation calculation is a forecast of net demand for oil produced by the cartel core—supplemented by work on core-country supply, which is part of the overall supply-studies effort. Together, these components form the basis for study of current market characteristics, and forecasting of possible future developments.

The estimation of price-taker supply is a critical aspect of this analysis, and it is to this topic that we now turn.

2. PROBLEMS OF SUPPLY ANALYSIS

Several types of analysis have been used to explain and forecast petroleum supply. One approach that has gained acceptance in recent years, particularly in the United States, is the use of econometrics.[3] This technique has been applied in circumstances where hundreds of large fields, each containing a number of reservoirs, have given the productive systems the stability of large numbers, and where the depletion effect (tending to raise costs as less of a reserve remains) has for a long time been offset by new discoveries and improvements in technology. Recently this balance seems to have been lost. Also, in the data used in the econometric studies there seems to have been some ambiguity about the meaning of reported "reserves" and changes therein, so that a given year's reported "discoveries" bore little relation to what had actually been found. Moreover, there was no explicit attention to costs, which might cause a given price to be profitable in one place but not in another.

In studying supply from many areas of the world, the conditions for econometric analysis are even less favorable than in the United States. In many countries the oil fields are both fewer and younger, and even the short histories are poorly documented. Another limitation is the fact that the price series are so fragmentary and untrustworthy. The so called "posted prices" of the past were rendered meaningless around 1960, when they became artifacts used for the calculation of taxes. Moreover, data on arms-length sales of crude oil are insufficient and are ridden with too many errors to serve as a basis for econometric investigation. Were those problems not serious enough, current prices are far outside the historical range, and the inputs needed for development are not necessarily available in easy supply, either at constant or predictably changing prices. Finally, econometric calculations assumed, correctly, the existence of a competitive industry in the United States, hence a competitive supply curve. But our task is the modeling of a cartel, where price changes may have perverse effects on output.

Another approach to supply forecasting, which also involves an orderly summation of the past, is that typified by the work of the National Petroleum Council (15), and subsequently applied by the U.S. Federal Energy Administration (20). Under this approach, the experience of past exploratory drilling is summarized and a trend in the finding of reserves per foot drilled is established. Based on estimates of the costs of exploration and development, calculations are made of the relative attractiveness of exploratory activity, conditioned on some assumption about the price of oil. Given an estimate of exploratory drilling, the forecast of barrels added per foot drilled, a reserve-to-production ratio, and hoped-for stability in reserve expansion in old "fully developed" fields, it is possible to forecast supply into the future.

Unfortunately, many of the shortcomings of the econometric approach apply as well to the NPC-type format. In many areas of the world the exploratory histories are poorly documented, and several of the relationships which are required for this approach may be estimated only very approximately. This is because many important supply areas of the world are relatively new, and the experience that makes the NPC method believable simply does not exist. Moreover, the more productive potential areas in the world often are located in offshore or otherwise inaccessible areas, and the cost of development-production of particular resources weighs very heavily in the supply relation, as opposed to the phenomenon of exploration and finding which is emphasized in the NPC method.

Finally, there are the methods of resource estimation used by oil companies in evaluating prospective areas, and in constructing global estimates of regional or world resources. These methods, which draw on detailed geologic and geophysical data as well as on past drilling experience, seem to be rarely used for supply estimation of the type being carried out here. Where they are so applied, it usually is not possible to gain access to the details of underlying data and assumptions. They do, nonetheless, contain important components of concept, information and analytic method, and use is made of these approaches below.

Key Factors in the Analysis

To a very great extent, therefore, the mechanisms we have chosen to use are determined by what we perceive as the severity of data limitations in the main variables. The first and fundamental problem is in the reserve data. Viewed as an economic process, oil supply is the depletion of a stock, which is constantly being renewed by adding new reservoirs and expanding the limits of the old ones. "Reserves" have for years been reported by The American Petroleum Institute (API) for the United States on a consistent and meaningful basis (1). Essentially they represent "money in the bank": in effect, the organized consensus of industry personnel as to the amount to be produced from existing installations. Elsewhere, one must rely on government estimates whose basis is rarely revealed, and on the trade press, which also is essentially an informal consensus of company opinion. The main difficulty in using non-API estimates results from various conceptions and definitions of what is likely to be added, how soon, to "proved reserves." In all cases, "reserves" are not a direct measurement but an inference drawn from data on geological structure plus observations on production volumes, pressures and temperatures. There will be legitimate differences of opinion in the interpretation of such data, hence in the estimation even of "proved" reserves. These variations are magnified as one moves from reserves proved to those considered as "probable" in the existing cluster of reservoirs.

A considerably greater leap, and the one calling for more specialized knowledge, is the estimation of "undiscovered reserves." Less than a decade ago, such estimations were an exercise in method; or in the language of a distinguished geologist, Lewis Weeks, they were merely an indication of where an exploration department ought to go look. These estimates had, in short, only a relative meaning, and it was a plain error to compare them with, or add them to, proved or probable reserves in known reservoirs. But today one can estimate the ultimate reserves for a "trend" or "play" (i.e., a population of reservoirs, generated by a geological sequence) provided that enough is already known about the area to furnish a reliable sample. Such a method is presented in section 3, below. The combination of mathematical statistics and geologists' knowledge is not easily created, however, and we have been able thus far to apply the method in only one area, the North Sea. A number of estimates have been made, much more approximately, for larger areas. These are discussed in section 4. Rather than rejecting or ignoring them as not good enough, we regard them as a considerable advance on simple extrapolation on the basis of cubic yards of sedimentary rocks, etc.

An equally important data limitation is from reservoir engineering: how much can be produced out of a given set of reservoirs in any given time. In the past few years these limits have been perceived as much more tightly binding. When the price of oil in the United States was around $3 per barrel there was little dissent from the view that if the price were $6, vast new reserves could be created by applying more capital and extracting much more than the average 30 percent of the oil in place. These hopes are not dead, but it is now seen that too little was known of the processes by which additional oil could be recovered from a given reservoir, in the field rather than the laboratory. In the United States, drilling has responded to price, but reservoirs have not.

There also has been some unpleasant learning about the amount which can be produced per day or per year without damaging the reservoir and lessening the ultimate recovery to the point where present value is also less. Iran is one example. We happen to have obtained the capital budgets of the Iranian Consortium for over a decade (6). Reading them in succession makes it plain that for years there was no felt need to know what would happen if production were raised by several times. It was reasonable to foresee, at prices much less than now rule at the Persian Gulf, capacity of 10 million barrels daily (mbd). The maximum will probably be a third to fourth less. Instead of a continuum, with higher prices bringing out higher output rates, the marginal cost appears to become nearly vertical in the neighborhood of 6 to 7 mbd. Given the strategic position of Iran among oil producing nations, this change has had important consequences. We can only be sure it is not the only one nor the last.

Finally, there is the influence of government policies. In competitive industries, supply and demand will be equated by price; in noncompetitive industries, by marginal revenue. World oil is a good deal more complex. At current prices, the margin of price over costs is very great even in the highest-cost areas. Where the industry is operated by private companies, payments to the government greatly exceed payments to factors, including capital charges. Hence the most important economic variable, sometimes by factors of 10 or even 100, is the government's perception of how great a rent exists, and how high a price they can charge without reducing their total take. But the government take may also be in the form of participation or joint control. There is much room for misunderstanding and deadlock, so that a given country's actual rate of development may be much below where it would be under a government which was better informed and free to maximize, without political or ideological pressures. Matters are simpler in those countries where the oil industry is owned entirely by the government. As a first approximation, given knowledge of development costs and of known and probable reserves, one can calculate the rate of output which would maximize the present value of the current reserves—as well as estimate the finding rate which maximizes the present value of reserves in reservoirs to be discovered in the known areas. But one may need to modify the approximation to accommodate cartel solidarity, or other objectives.

In some countries, there is a backward-bending supply curve, where higher prices lead to less supply. A government may simply overreach itself, to take such different cases as Canada and Malaysia. Higher prices may promise to generate so much revenue as to disrupt the desired rate of social adaptation; as a result, the higher the price, the less the target rate. A government with certain plans or obligations can meet them, given higher prices, with less output; hence is willing to reduce output or at least to accept reductions. Finally, price increases always generate expectations of still further increases. The raises the present perceived value of any reserves, and lowers the optimal rate of development.

Thus the three basic determinants—reserves, development costs, and government policies—must be put into a framework where they can be acted upon by current and expected prices. The framework must be modular to an extreme degree, since there is hardly a piece of the data base which we may not need to replace at any time, as more becomes known, or as data become outmoded.

Two Approaches to Price-Taker Supply

Here we present two of the analysis methods that we are exploring. One of these, the "disaggregated pool analysis," is the most detailed of the models developed and requires the most data, geological interpretation, and computational capacity. The other, which we call "aggregated country analysis," is among the most simple of the methods formulated. Various extensions, modifications, and combinations of the two approaches remain to be explored.

Though one approach is far more ambitious than the other, essentially they are variations of the same model. That is, the supply function is based on a simulation of the process of exploration, development, and production within a petroleum region. Both take account of geological data, though one has a more complex hypothesis about geological deposition and the nature of the exploratory process. Both take account of economic factors such as costs and future petroleum prices, and both include an approximation of oil-developer decisions, though the aggregated method necessarily treats these in a highly summarized manner. Both allow for the effects of tax regimes and other aspects of producer country oil policy, though once again the details of tax structure must be sacrificed in the aggregated model.

Figure 2 shows how the two analyses fit into the sequence of activities which compose the oil-production industry, and the kinds of statistics generated. Looking first at the left-hand column, we see an immense store of experience, combined with formal science and technology, which gives rise to judgmental estimates of what may be contained, and is worth producing, in various parts of the world. The unknown areas are judged by analogy with the known. The kinds of pools which may be generated, and the relative size distributions, constitute the estimates of "ultimate production," as compiled by several oil companies.

These estimates have a direct effect on the direction of geological and geophysical prospecting. The knowledge gained thereby feeds back into the judgmental estimates. Geological-geophysical results also determine exploratory drilling, in new and in old areas; and the good or bad results again feed into judgmental estimates, both of "ultimate production," and of what may be thought, in old areas, of likely new discoveries there, i.e., "probable reserves."

Exploratory drilling generates dry holes, which are not necessarily bad news, since the information about successive layers may be of great value; it also leads to new fields in old areas, again affecting estimates of "probable reserves" and, perhaps more weakly, of "ultimate production."

At this point the Disaggregated Pool Analysis can begin. Given enough wells drilled to furnish a reliable sample, and given geological knowledge to certify the existence of a population from which the sample is drawn, a forecast is made of the underlying distribution, the number of reservoirs to be found, and their size distribution. Given knowledge of costs, prices, and tax policies, one can make a forecast of the rate of exploratory drilling, and the rate of development of the area.

Stated in terms of the steps of our analysis procedure, the disaggregated

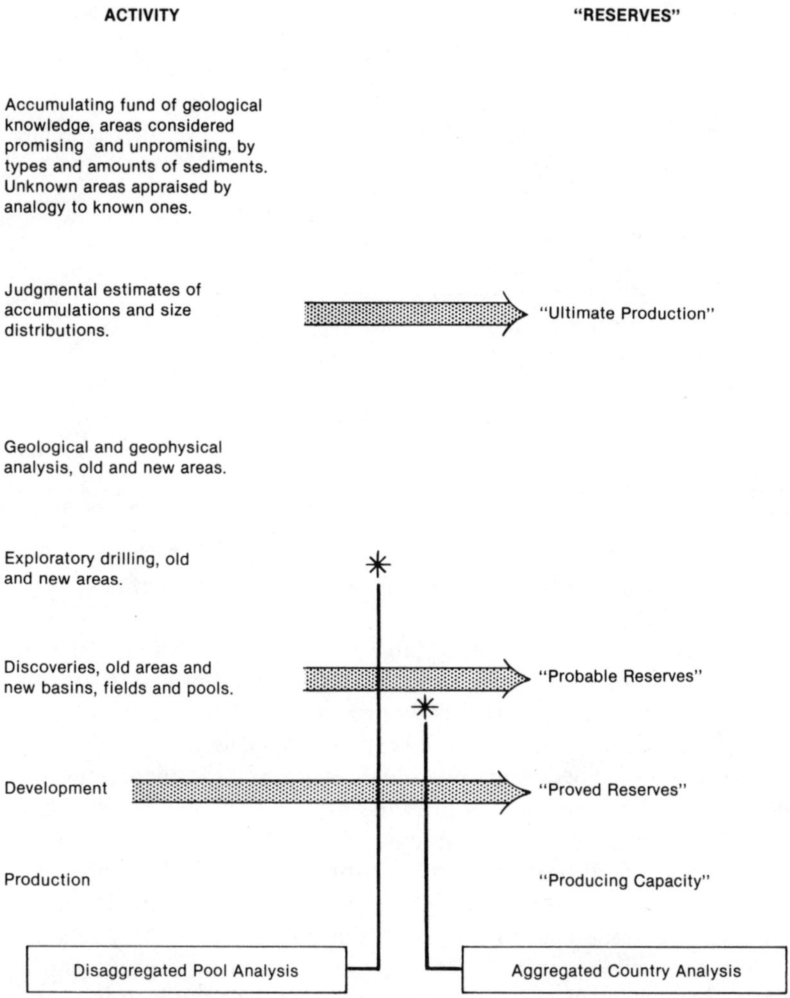

Figure 2. Sequence of oil activities and kinds of statistics generated.

pool approach is shown in the left half of Figure 3. First, an estimate is made of the number of exploratory wells in the region. Then the exploratory process itself must be approximated, and an estimate made of the number of reservoirs found and their characteristics—e.g., recoverable reserves, well productivity, depth to pay, and (in the case of offshore areas) water depth and distance to shore. The key variable is recoverable reserves, and here the method draws on research on statistical analysis of the exploratory process carried out by Kaufman and Barouch (2, 3). The economics are then calculated, reservoir by reservoir. Development costs are estimated from the

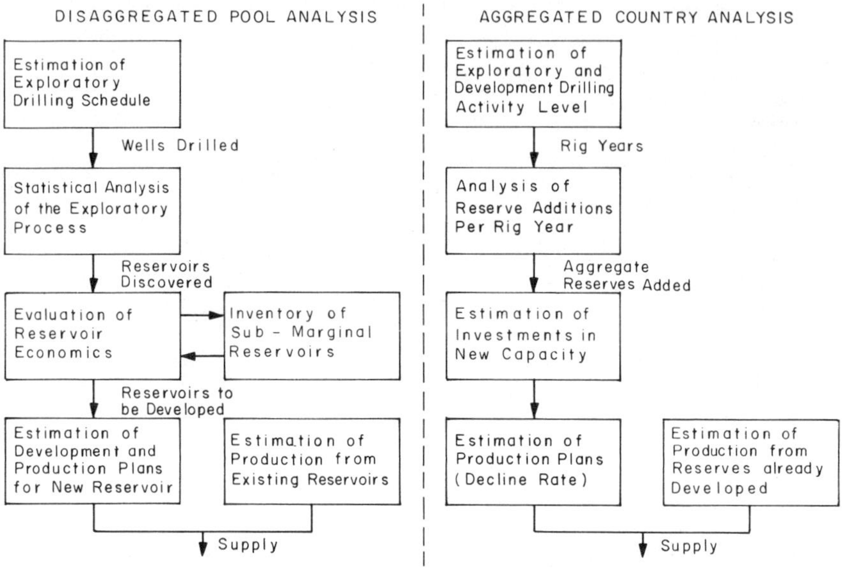

Figure 3. Outline of alternative supply analysis methods.

reservoir characteristics and applicable tax rules, and these are combined with an evaluation of revenues based on expected oil prices. Built into the analysis is an evaluation of the likely production profile for reservoirs with different characteristics. The overall supply estimate for an area is the sum of the production profiles of its individual reservoirs; and the overall supply from a country (or other region) is the aggregate of the results from the various areas distinguished for analysis.

The disaggregated pool analysis is a formal, precise, reproducible version of what happens in industry every time a discovery is made. Estimates of probable reserves are changed, as well as of "ultimate production." As development wells are drilled, providing new producing capacity and more detailed knowledge of reservoirs, the forecast of their production is crystallized into the industry estimates of "proved reserves."

The aggregated analysis is an attempt to tap into the calculations which the industry has already made, which are in substance the same as the disaggregated pool analysis. By considering prices, costs, and taxes, the companies have already made estimates of what they expect to produce. Recent experience indicates how fast a given amount of drilling effort will produce the associated productive capacity, i.e., the ability to deplete a given reservoir in a given length of time. And the "probable reserves" and "ultimate production," both affected by recent experience, are imprecise measures of the larger "pool" out of which are being impounded the newly "proved" reserves. The smaller the proportion impounded up to the present, the more is left to be taken into "proved reserves" before rising marginal costs are felt. This is strictly analagous to the discovery decline rate which is made explicit in the statistical analysis at the pool level.

Once again, the approach can be expressed in terms of the explicit steps to be worked out in section 4. As indicated in the right-hand panel of Figure 3, exploratory and development activity are measured in rig years for the area in question, and an estimate is made of proved reserves added per rig year. Instead of making a detailed study of reservoir economics, we factor in the industry's estimates, since *all* proved reserves are by definition slated to be developed. The focus is on the amount of capacity that will be installed, considering costs and expected oil prices. The cost data are not in the form of cost functions, as required by the disaggregated model, but simple estimates for each country of the capital coefficients per daily barrel of capacity and of operating cost per barrel of production. As with the disaggregated analysis, separate estimates are made of supply from reserves *already* in production, as shown in the lower part of the diagram. And, of course, the method is flexible enough to take account of information about other aspects of planned activity.

Both methods are based on the same idea of how reservoirs are distributed in the earth and on the same sequence of industry operations. Whether we use one or the other depends altogether on the available data on the one side, and the money and manpower available to use them, on the other.

3. DISAGGREGATED POOL ANALYSIS

As noted above, the key to the disaggregated method of forecasting is the statistical analysis of the exploratory process, and economic evaluation of the pools that are found.[4] Methods have been developed whereby a forecast can be made not only of the total recoverable reserves to be discovered by a given level of exploratory effort, but of the distribution of pool size itself and of the sequence of discoveries by size. Pool size is a key to the economics of supply, particularly in offshore areas (or under certain tax regimes) where small pools (and under some conditions relatively large ones) may be uneconomic to develop. Pool size also is a significant influence on the speed of development and the time profile of production from reserves that are economic to produce.

Analysis of the Exploratory Process

As currently implemented, the analysis begins with a postulated level of exploratory effort within an oil "play." A play is defined as the pre-drilling and drilling exploration of a geological configuration, generated by a series of geological events, and conceived or proved to contain hydrocarbons.

The wells drilled form a statistical population. An important element of the research is to discover the difference in results that may accompany alternative ways of drawing these geographical boundaries. For example, the North Sea results shown here treat the entire region as a single play, but work is under way to perform the same analysis with the area divided into four areas that correspond more closely to the geologist's definition of a "play."

Exploratory effort is measured in terms of the number of exploratory wells, W_t, to be sunk in period t. To this drilling effort we apply a "dry hole risk," δ_t, which is the probability that an exploratory well will fail to find a field. For purposes of the example developed here, δ_t, is held constant over the forecast period, and reflects a judgmental estimate based on historical experience. The expected number of discoveries in a period can then be expressed as $W_t(1 - \delta_t)$. The purpose of the analysis of the exploratory process is to determine the expected characteristics of the next set of discoveries in a petroleum play over some period to be analyzed. This is the first step in calculating the contribution that new pools may make to future supply, as shown in Figure 3. This estimate can be combined with data on pools already discovered to produce an overall supply forecast for the play.

The analysis of the exploratory process is based on two hypotheses about the natural process of resource deposition. Following the work of Barouch and Kaufman (2, 3), onc assumcs that the number of pools in a play, N, and their size distribution (characterized by the mean and variance, μ and σ^2) are generated according to a probability law whose functional form is dictated by the way in which nature deposited the oil in the first place. Many distributions may be analyzed using the methods applied here, but the customary assumption is that pool size in terms of recoverable reserves, r, is a random variable, and that its density function normal; i.e., log r is normally distributed with mean μ and variance σ^2.

To this hypothesis about nature, then, is added another hypothesis about the process by which oil operators search for and find these reservoirs. Once again following the work of Kaufman and others (2, 3) the exploratory process is characterized as one of random sampling, without replacement, in proportion to pool size, r. With these hypotheses it is possible to predict the characteristics of discoveries number $n + 1, n + 2, \ldots, N$—and also total recoverable reserves in the play, $\sum_{j=1}^{N} r_j$—if we know there have been n discoveries of size $r_1, \ldots, r_n$ and the order in which they were discovered.

The procedure is as follows. We can specify joint distribution function of the n discoveries to date conditional on the parameters μ, σ^2, and N:

$$D(r_1, \ldots, r_n | \mu, \sigma^2, N). \tag{1}$$

Then, using the actual sequence, $r_1, \ldots, r_n$ as one sample from this distri-

Table 1. Northern North Sea Discoveries, Recoverable Reserves. Oil equivalent (millions of barrels)

Order	*Name of location*	*Date*	*Size*	*Order*	*Name of location*	*Date*	*Size*
1	Cod	2/68	156	31	Ninian	9/73	1,000
2	Montrose	4/69	180	32	Statfjord	12/73	4,960
3	Ekofisk	9/69	1,713	33	Odin	12/73	178
4	Josephine	6/70	100	34	Bruce	3/74	450
5	Tor	8/70	243	35	Magnus	4/74	800
6	Eldfisk	8/70	910	36	N. E. Frigg	4/74	71
7	Forties	8/70	1,800	37	Balder	4/74	100
8	W. Ekofisk	8/70	490	38	Andrew	4/74	300
9	Auk	9/70	50	39	Claymore	4/74	375
10	Frigg	4/71	1,325	40	E. Magnus	6/74	250
11	Brent	5/71	2,500	41	9/13–4	6/74	220
12	Argyll	6/71	70	42	15/6–1	9/74	150
13	Bream	12/71	75	43	Brae	9/74	800
14	Lomond	2/72	500	44	Sleipner	9/74	50
15	S. E. Tor	4/72	34	45	Hod	11/74	75
16	Beryl	5/72	525	46	211/27–3	11/74	450
17	Cormorant	6/72	165	47	Gudrun	11/74	450
18	Edda	6/72	98	48	2/10–1	11/74	100
19	Heimdal	7/72	300	49	3/4–4	12/74	100
20	Albuskjell	7/72	357	50	14/20–1	1/75	75
21	Thistle	7/72	375	51	Crawford	1/75	150
22	Piper	11/72	638	52	9/13–7	1/75	350
23	Maureen	11/72	500	53	3/8–3	1/75	100
24	Dunlin	4/73	435	54	Tern	2/75	175
25	3/15–2	4/73	150	55	21/2–1	2/75	175
26	Hutton	7/73	250	56	3/2–1A	3/75	200
27	Alwyn	7/73	350	57	Valhalla	4/75	50
28	E. Frigg	8/73	623	58	3/4–6 & 3/9–1		200
29	Heather	8/73	150	59	15/13–2		200
30	Brisling	8/73	75	60	211/26–4		175

Source: Beall (4), and estimates by the MIT World Oil Project as of June 1976.

bution we can estimate (using maximum likelihood techniques) the parameters μ, σ^2 and N. Table 1 presents the sequence of discoveries in the North Sea, which we use as the basis of the sample calculations presented below.[5]

Next, it is possible to specify the density function of discovery number n + 1 conditional upon the exploratory history already experienced, $r_1, \ldots, r_n$, and given estimates of the parameters $\hat{\mu}, \hat{\sigma}^2$, and $\hat{N}$. This we may write, defining $r = (r_1, \ldots, r_n)$, as

$$D(r_{n+1} | r; \hat{\mu}, \hat{\sigma}^2, \hat{N}). \tag{2}$$

An estimate of the expectation of r_{n+1}, the $(n + 1)^{st}$ discovery, given that r has been observed, is

$$E(r_{n+1} | r) = \int_0^\infty x D(x | r; \hat{\mu}, \hat{\sigma}^2, \hat{N}) dx. \tag{3}$$

The nature of this calculation can be seen in Figure 4, based on data from the North Sea. When this analysis was done there had been 60 discoveries in the North Sea, so n = 60. The figure shows the rough shape of the density function for the 61st discovery along with the conditional expectation

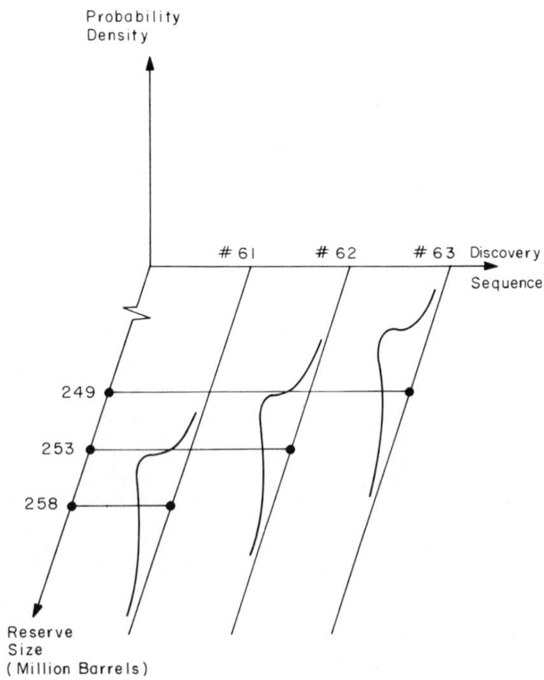

Figure 4. The sequence of predictive discovery distributions.

$E(r_{61}|r)$ of the size of the 61st discovery, which in this case was 258 million barrels. The calculation for n + 2, n + 3, etc. is a straightforward extension of this procedure.

One further step is necessary before proceeding to economic and financial analysis of the reservoirs themselves. As stated earlier, smaller pools may be uneconomic to develop, and the pace of extraction may differ among larger pools depending on size. Therefore we need some indication of the expected number of barrels to be found in pools of various sizes. The data for such a calculation is contained in the conditional distribution of Equation (3), and using this function we can calculate the partial expectations of the numbers of barrels in reservoirs of various sizes. To do this we define a set of k class sizes S_k, where the size limits are defined as shown in the following table.

Class (k)	Lower Limit	Upper Limit
1	a_1	a_2
2	a_2	a_3
3	a_3	a_4
.		
.		
.		

Then for the $(n + 1)^{st}$ discovery the expected number of barrels to be found in pools in size class k, or the partial expectation of size class k, is

$$P_{n+1,k} = \int_{a_k}^{a_{k+1}} xD(x|r;\hat{\mu},\hat{\sigma}^2,\hat{N})dx. \tag{4}$$

The results of this calculation are illustrated in Table 2, once again using data from our North Sea example. Four size categories are used and the

Table 2. Predictive discovery distribution (Millions of barrels oil equivalent)

Size category, k	Limits	Partial expectation. P_{ik}, for discovery number 61	62	63	64	65
1	125 to 250	18	18	18	17	17
2	250 to 375	26	25	25	25	24
3	375 to 500	31	31	30	30	30
4	over 500	176	173	169	166	163
Expected value, $E(r_i)$		258	253	249	244	240

table shows the partial expectations of the number of barrels to be discovered in each category in the next five successful exploratory wells.

Though the table shows only the first few discoveries from a longer sequence that must be generated for supply forecasting, several characteristics of the process are evident in the data shown. First, most of the oil is expected to be found in larger reservoirs, and the difference in *economic* reserves which would result were the smaller-size reservoirs infeasible to develop is not great—though it is significant. Second, the table plainly shows a process which we refer to as "discovery decline." That is, as the province is drilled up, the expected finding from each additional successful exploratory well tends to decrease. This is the behavior one would expect in practice, and it falls out of the analysis because the fundamental geological facts of life are built into the method through the two key hypotheses introduced earlier.[6]

Referring again to Figure 3, the results in Table 2 constitute the data on "reservoirs discovered." The next step, then, is to estimate what will be done with them once their location and size are known.

Evaluation of Reservoir Economics

There are a number of attributes of an oil reservoir that influence its economics. The most important is recoverable reserves,[7] and this we predict by the methods above. Reservoir depth and (in offshore areas) water depth and distance to shore also are important determinants of cost—though these usually do not vary significantly over a region of the size on which this analysis is based. Likewise, average well productivity is an important factor, though it also is reasonably assumed to be constant over the unit of analysis when a relatively disaggregated approach is taken. These attributes—together with cost factors, tax rules, and the expected oil price—determine the economic viability of a reservoir.

We are concerned here with the behavior of price-taker suppliers, and in particular with the forecasting of their response to the price strategy of the cartel core. Since smaller reservoirs are more expensive to develop, per barrel produced, one may expect that higher cartel prices will bring more resources to the point of economic viability and call forth increased price-taker supply. To analyze this phenomenon, we construct a cash-flow analysis of the reservoir from the operator's viewpoint.

A profile of capital expenditures typical of oil reservoir development is shown by the solid line E_t in the top half of Figure 5. We have prepared an analysis of reservoir economics to generate estimates of E_t for reservoirs of different size (9, 19). Relationships were estimated for the various components of capital cost (e.g., development drilling, platform structures, platform equipment, pipelines, terminals) and of operating cost. In this analysis of the North Sea, extensive use was made of data prepared by Wood, Mackenzie,

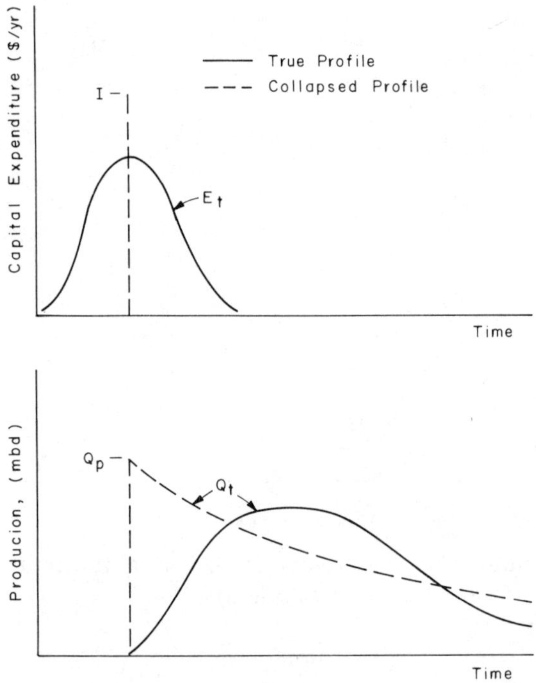

Figure 5. Profiles of capital expenditure and production.

and Co. In addition, detailed consideration was given to the tax systems of the United Kingdom and Norway.

Estimates were also prepared of the typical patterns of annual production, Q_t, from reservoirs of different size, as shown by the solid curve in the bottom half of Figure 5. These data are then combined into a cash-flow analysis of reservoir investment and production, both to determine economic viability and to establish the likely production profile, reservoir by reservoir. (The dashed lines in Figure 5 represent a simplified or "collapsed" version of development expenditures and the associated production profile. The collapsed model is used in constructing the aggregated model of oil supply, discussed in section 4.)

The result of the reservoir analysis is shown in Figure 6, once again using data from the North Sea. The figure shows a plot of the minimum-sized pool which it is feasible to develop given an expected level of oil price. At $6 per barrel (1976 prices), no reservoir below 200 million barrels will be developed given current costs and tax rules in the North Sea. At a $12 price, the marginal reservoir decreases to 90 million barrels. Similarly, it is possible to hold price constant and calculate the effect of changing tax rules.

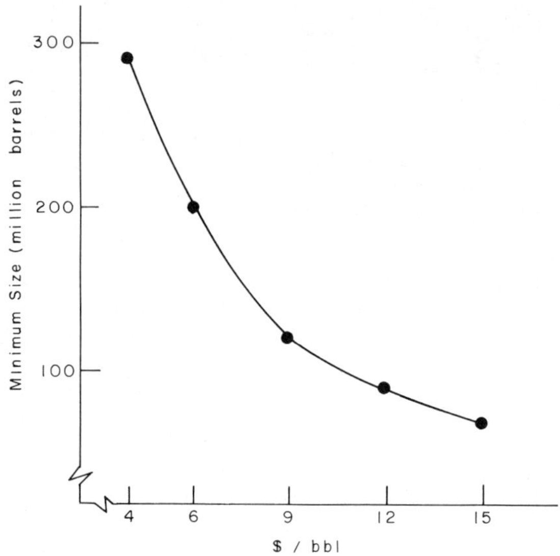

Figure 6. Minimum economic reservoir size as a function of oil price (1976 prices).

Simulation of Supply

The analysis of the exploratory process, presented in Table 2, and the evaluation of reservoir economics, summarized by Figure 6, then constitute the building blocks for simulation analysis of price-taker supply. Refer again to the left-hand side of Figure 3. Based on an estimate of exploration drilling activity (which for this example we estimate from announced plans) a discovery sequence is generated. It represents the expected value of the reserves to be found in pools of various sizes. These expected quantities are then subjected to analysis of economic viability. Submarginal reservoirs are held aside until such time as rising oil prices or postulated revisions in tax rules change their attractiveness for development. Those reserves that are in pools above the minimum pool size are then converted (based on analysis of reservoir economics) into a time profile of production. To the estimates of supply from new discoveries, then, are added data on expected production from reservoirs already in production or under development. These, of course, constitute the more accurate portion of any forecast. In the North Sea we depend heavily on Wood, Mackenzie data for these estimates.

The final result is a supply forecast built up pool by pool for existing fields and (on an expected value basis) discovery by discovery for new reservoirs. If there is more than one play in the supply region, then the regional production is built up play by play. The resulting supply estimates for the North Sea are shown in Table 3, and compared with estimates of other analysts.

Table 3. North sea supply estimates (million barrels per day, oil equivalent)

	1980		*1985*	
	$*9*	$*12*	$*9*	$*12*
Our sample analysis[a]				
Existing reservoirs	2.82	2.82	2.50	2.50
Recent discoveries	1.68	1.82	1.57	1.66
1977–1978 discoveries	0.47	0.48	1.68	1.72
Total	4.97	5.13	5.75	5.88
Energy Prospects to 1985 (17)[b]	3.9	4.06	5.16	5.3
Birks (5)[c]	3.46		6.8	
Odell and Rosing (16)[c]	4		12	

a. 1976 Prices.
b. 1974. Based on 1972 prices of $6 and $9 in the Persian Gulf, which are used here as proxies for 1976 prices of $9 and $12 in the North Sea.
c. Price assumption not specified.

4. AGGREGATED COUNTRY ANALYSIS

It is only in special circumstances where we have the data needed for the disaggregated analysis, so we are developing a greatly simplified, but consistent, aggregated method sketched in Figure 3. It is modular, in that the estimates for any given country, region, or field can be replaced by more precise knowledge when available from other sources.

As noted earlier, the method is based on forecasts of rig activity and analysis of proved-reserves added per rig year. Reserve additions then become an input to calculations of capacity expansion and likely oil production. Therefore we turn first to the data sources for world-wide reserves, and the way they may be used to forecast new additions.

Analysis of Reserve Additions

As used by the American Petroleum Institute (API) the concept of *proved reserves* has a definite economic meaning: *a highly accurate forecast of what will be produced* from wells and facilities already installed. Since variable costs are normally only a small part of the total, it would take an unusually severe price drop to abort much production. It is this concept of reserves that forms the basis of the model developed below.

However, outside the United States published "proved reserve" estimates generally include a substantial element of what the API calls "indicated additional reserves from known reservoirs." At end-1975 in the United

Table 4. Oil company estimates of proved reserves (API concept) as a Percentage of published reserves (*Oil and Gas Journal*), 1970.

Area	*Proved reserves*	*Proved plus probable reserves*
Latin America	.97 to .99	Not available
Europe	.97 to .98	
Africa	.50 to .73	
Middle East	.67 to. 80	
Total	.66 to .81	.88 to .97
Rough point estimate	about .75	about .95

Source: National Petroleum Council Committee on U.S. Energy Outlook: an Interim Report. An Interim Appraisal by the Oil Supply Task Group, 1972, pp. 21–24.

States (excluding Alaska) this category included 5.0 billion barrels compared with 22.7 billion proved reserves, or an additional 22 percent (1). Usually the published "proved reserves" go farther. Interesting data were gathered in a survey of oil companies conducted by the National Petroleum Council. The companies made estimates of the discrepancies between API "proved reserves" and the reserves as published by the *Oil and Gas Journal* (*OGJ*). The results are shown in Table 4. Since the North Sea was insignificant in 1970, the close agreement on Europe is now outdated. Essentially, *OGJ* reserves for certain large Persian Gulf and African countries include a large amount of oil not yet developed into proved reserves, and the companies were not unanimous on its size.

Table 5 Reserve estimates end-1975 for selected countries (billions of barrels)

	1	*2*	*3*	*4*	*5*
		IPE Identified Largest Fields			
Country	*OGJ Published Total Reserves*	*No. of Fields*	*Reserves*	*Total Estimated Proved Reserves*	*Column* (4) *as Percent of Column* (1)
Abu Dhabi	29.5	4	4.8	6.4	21.5
Iraq	34.3	6	24.7	27.4	79.9
Kuwait	68.0	all (8)	54.6	52.2	76.8
Saudi Arabia	148.6	12	112.6	118.1	79.5
Iran	64.5	17	49.6	52.5	81.3
Nigeria	20.2	12	3.4	10.6	52.3
Libya	26.1	13	21.3	23.0	88.1
Venezuela	17.7	21	14.7	16.4	92.7

We gain a different and more useful perspective on reserve estimation by noting that reserves credited to individual developed fields fall substantially below the *OGJ* national totals. Table 5 shows a comparison of *OGJ* data and estimates of the largest fields as published by the *International Petroleum Encyclopedia* (*IPE*), along with an estimate based on the *IPE* figures. The estimate of proved reserves (column 4) was prepared by dividing *IPE* reserves of identified largest fields by the portion of total country production accounted for by those fields. Finds accounting for less than 3 percent of current production were excluded. (In Libya the proportion was based on cumulative not current production.) These authors' estimates as a percentage of the *OGJ* figure is shown in column 5; the results are consistent with those in Table 4.

Prior to 1975, the percentages in column 5 would have been lower, since *OGJ* in that year revised its estimates downward, and for the first time labeled them "proved reserves" and stated that "probable and possible reserves [are] not included." Given this definition, we should consider that the excess of *OGJ* reserves over those estimated on an *IPE* basis includes largely the undeveloped portions of known fields, including some known but undeveloped reservoirs. For our purposes, the *IPE*-based figures are preferred for they are closer to the API definition, though considerable judgment may be involved in making estimates for particular countries.

Given an estimate of current proved reserves, we proceed to develop a method for forecasting this quantity. We consider gross additions to proved reserves in any year as an output, and rig-years as a proxy or indicator of investment input which generates these reserves. We let R_t be proved reserves in an area in year t, Q_t be the area's production, and RY_t be rigs operating there. Then reserves-added per rig time unit, RA, is calculated as

$$RA = \left[R_{75} - R_{72} + \sum_{t=73}^{75} Q_t \right] \sum_{t=73}^{75} RY_t. \tag{5}$$

The numerator is gross additions to proved reserves over the three-year period, 1972–1975, where additions include new-field discoveries, new pool discoveries, and revisions and extensions of known fields, often from development drilling. The denominator is the number of rig years during the same period. Rig time is superior to feet drilled because it is a better predictor of investment, although it is necessary to calculate RA separately for onshore and offshore areas. Rig time includes all time-related elements, including not only depth but also time used in moving rigs[8]; interruptions for lack of an essential part or service or any other reason; unusually different drilling conditions, etc.

Equation (5) applied to a subsequent year's drilling rate yields a forecast of reserve additions:

$$\Delta R_t = RY_t \cdot RA. \tag{6}$$

The method is a simple extrapolation of recent experience, with only inertia (which is considerable) to justify its use.

We need to take care to avoid some obvious inaccuracies. Thus, for example, we disregard the 1973–1974 Venezuelan "marking up" of proved reserves to the extent of about 4.5 billion barrels.[9] The validity of the changes seems questionable. Elsewhere, no price effect is perceptible. More particularly, in the United States the recovery factor was not significantly different in 1975 from 1972.[10] It would appear that the effect of higher prices on supply makes itself felt only by increasing investment, i.e., drilling and reserves-added.

Of course, the yield from new investment cannot go on forever, undiminished by the effects of depletion. Thus we define a coefficient b_t to reflect this phenomenon. Let Cum R_t encompass all past production plus current proved reserves in year t. Ultimate recoverable reserves, Ult R_t, are then Cum R_t *plus* all future additions to reserves (and, ultimately, to production). If 1975 is the most recent year when reserves data are available, then b_t may be defined as

$$b_t - \left[1 - \frac{\text{Cum}\,R_t}{\text{Ult}\,R_t}\right] \left[1 - \frac{\text{Cum}\,R_{75}}{\text{Ult}\,R_{75}}\right]. \tag{7}$$

And the expression for reserve additions is modified to take this factor into account:

$$\Delta R_t = RY_t \cdot RA \cdot b_t. \tag{6'}$$

As new reserves are created by drilling, they are essentially a transfer out of the pool of ultimate production, Ult R_t, for the country or area. Thus, if cumulative production plus the amount already impounded into reserves at any moment were the same as the ultimate production, then the numerator of equation (7) would be unity, b_t would be zero, and no amount of drilling could add anything to reserves. The closer the numerator is to zero, the smaller is the fraction, and therefore the less is the return to drilling effort, relative to the 1973–1975 showing.[11] If, and as, the ultimate reserve estimates are changed up or down, or if we are confronted with varying estimates of ultimate reserves, we substitute them into equation (7) and see what difference it makes.

This aggregated method, with national entities as building blocks, treats as one reservoir what may be a collection of hundreds. Small countries may be left that way. But larger ones must be, as soon as possible, divided into rational subgroups. (In every case, we divide between onshore and offshore.) We can illustrate the method using the important new Reforma area of

Table 6. Sample calculation of reserves added for Reforma fields, Mexico.

a. Reserves added, 1973–1975	5.7 bb*
b. Development rig years, 1973–1975	99**
c. Reserves added per rig year [RA]†	0.058 bb
d. Ultimate recoverable reserves [Ult R_{75}]	60.0 bb††
e. Cumulative production plus proved reserves, Reforma only, end–1975 [Cum R_{75}]	5.7 bb‡

Yr.	*Develop. rigs*	*Reserves-added decline coefficient, b_t*	*Proved reserves added, RA_t (bb)*	*Cumulative proved reserves added (bb)*	*Ttl. Annual production capacity‡‡ (mbd) Depletion rate* 8%	5%	*Actual production§ (mbd)*
1974	33	1.000	1.91	1.91	0.121	0.077	0.113
1975	33	0.965	1.85	3.76	0.349	0.224	0.288
1976	33	0.931	1.78	5.54	0.713	0.461	0.432
1977	33	0.898	1.72	7.26	1.030	0.677	N/A
1978	33	0.866	1.66	8.92	1.320	0.875	
1979	33	0.835	1.60	10.52	1.570	1.050	
1980	33	0.806	1.54	12.06	1.790	1.220	
1981	33	0.778	1.49	13.55	1.970	1.360	
1982	33	0.750	1.44	14.99	2.140	1.500	
1983	33	0.724	1.38	16.37	2.280	1.620	
1984	33	0.698	1.34	17.71	2.390	1.720	
1985	33	0.673	1.29	19.00	2.490	1.820	

**International Petroleum Encyclopedia* (*IPE*), 1976, p. 222.

**Petroleos Mexicanos, Report of Director General (March 18, 1974 and 1975), Chiapas-Tabasco only. According to Hughes Tool Co. report, there was no substantial increase of rigs in the South Zone from 1974 to 1975. Hence, we assume the same number of development rigs throughout.

†Line (a) divided by line (b); see equation 5.

††Minister of National Patrimony, Francisco Javier Alejo, quoted in *El Tiempo*, May 17, 1976, and *Petroleum Economist*, June 1976, as expecting 7 million mbd by A.D. 2000. A linear growth, and assumption of 10 percent depletion rate in A.D. 2000 indicates ultimate added reserves of just over 60 bbl.

‡Original Reforma reserves. Differs from line (a) by 100 million bbl already produced.

‡‡Assumes: (i) a lag structure whereby new proved reserves are fully developed at the following rates: same year (t) 30 percent, year (t + 1) 30 percent, year (t + 2) 40 percent. This estimation is based on observed lags to full capacity in publicly-reported proved reserves. And (ii) either (a) 8 percent depletion annually of pre-existing and of new capacity, which is a weighted average of all large producing Mexico fields excluding Reforma (as calculated from *IPE*), or (b) 5 percent depletion annually, which is the announced objective. See *IPE*, op. cit., p. 191.

§ Reforma fields average production for the first six months of each year (calculated from *OGJ*). *IPE* reports total year averages of 0.0 mbd in 1974, 0.274 mbd in 1975, and 0.553 mbd in 1976. The estimates for 1974 and 1975 are probably inaccurate. The Prospectus for Mexican External Bonds, September 1976, is presumably authoritative, and reports 0.400 mbd average for the month of December 1975. The 1974 *PEMEX: Memoria De Labores*, p. 13, reports 0.171 mbd as the total 1974 average with a rate of 0.275 mbd reached in December 1974.

Mexico.[12] In this region, the new reserves—proved, probable and ultimate—have all been added since 1972, and therefore we treat the area separately from the rest of Mexico. To apply to the newly developing areas the coefficients derived from areas a half century old or more would have been right only by chance.

The Reforma results are shown in Table 6. The top part of the table shows the key parameters needed for equations (5)–(7); the bottom portion presents the results. Column 5 shows cumulative reserves added totalling 19 billion barrels (bbl) by 1985. Also shown in the table is a forecast of annual production (to be discussed in the following section) which is based on this forecast of reserves-added. The same analysis performed for the rest of Mexico, or for any other nation or subnational region, will look very much like Table 6.

In this example, an assumed constant rate of development drilling, interacting with government estimates of proved reserves and ultimate recovery, yields a peak production rate of 2.39 million barrels per day (mbd) reached in 1984, declining thereafter. Yet high-ranking officials forcast a production rate of 7 mbd in the year 2000. This implies considerably larger reserves-added, through more drilling activity, or more effective drilling (reduced rig time per well) than assumed in the model. Note we have assumed a *constant* number of development rigs. More development drilling would mean proportionately more production, lagged about three years. Thus we have identified one or two key variables in Mexico's plan which will change, and to which we must adjust our data set accordingly.

Analysis of New Capacity and Production Plans

We need now to go from proved reserves-added to attainable new capacity, as indicated in Figure 3. In some areas we have the companies' own plans. For actual forecasting, they dominate any estimates we could make. Moreover, they are a valuable check on our own calculations, which we make by two possible methods. We may rely again on the inertia of the system, and base production forecasts on the historical relation of output to proved reserves. We show this method first. Or, it may be possible to analyze the economic forces underlying observed production behavior, and methods of doing that also are discussed below.

But whichever method we use, application of a single depletion rate to a reserve estimate—to obtain a production profile—is a strong simplification, as Figure 5 shows. In the lower part of the figure, the solid line shows a typical production profile for a field; the area under the curve is the total proved reserves R. The dashed line is our simplified version of oil exploitation where production jumps immediately to an initial (and presumed peak) capacity Q_p and declines at a constant percentage rate thereafter. Once again, the area under the curve is R, so that

$$R = Q_p \int_0^T e^{-at} dt. \tag{8}$$

As $T \to \infty, R \to Q_p/a$; or in the limit, $a = Q_p/R$. The question for analysis then is: what is the value of the depletion rate, a (and therefore of Q_p), that is appropriate for new additions to reserves? Once a is determined it is a simple step to forecasts of supply. The remaining issue is the lag between the time proved reserves are "booked" and the point when the new capacity Q_p is on line. In a mature producing area the lag from the former to the latter is short, but not in a new province. We use, tentatively, two years onshore and four years offshore, but in every case including only those fields where development drilling has been started.[13]

Historical Production-Reserve Ratio. As a first approximation, capacity plans can be calculated on the assumption that new additions to proved reserves will be depleted at the same rate as existing fields. Under the assumption of a uniform policy regarding depletion, the depletion rate for an area can be calculated as

$$a = \sum_{t=1973}^{1975} Q_t/R_{73-75}, \tag{9}$$

where 1973–1975 is taken as a reasonable base period for estimation, and R_{73-75} is the average level of proved reserves over that period. Then for any new addition to proved reserves in year t, the new installed capacity is

$$Q_{pt} = a\Delta R_t. \tag{10}$$

Assuming no excess capacity is installed, production from R_t begins at the level $Q_t = Q_{pt}$ and declines at a percent per year.

However, the production-reserve ratio for any country is an aggregate of many fields. Hence a simple division of national production by national reserves may be seriously misleading. A good example is Table 7, showing Abu Dhabi. Total national production was 1.7 percent of proved-plus-probable reserves. But the bulk of those "reserves" are still undeveloped; the working inventory is being drawn down much faster. If we confine ourselves to proved reserves, the simple quotient $Q/R = .08$ gives equal weight to every barrel of developed reserves. But our true objective is to give equal weight to every barrel produced. If we want to estimate how much Abu Dhabi is capable of producing next year, the Bu Hasa field (156 million bbl) is approximately twice as important as Zakum (82 million).

Accordingly, we weight the production-reserve ratio for each field by its production, and divide the total for all fields by the total weight. This calculation is shown in Table 7. Use of the weighted average mean lets us

Table 7. Weighted mean depletion rate, Abu Dhabi, 1975 (millions of barrels)

Field, Discovery Date	*Production* Q	*Reserves* R	Q/R	Q^2/R
Asab, 1965	90	500	.180	16.2
Bu Hasa, 1962	156	1,289	.121	18.9
Mubarras, 1971	7	150	.047	0.3
Umm Shaif, 1958	59	1,706	.035	2.1
Zakum, 1964	82	1,314	.062	5.1
Subtotal	394	4,959	.080	42.6

Unweighted mean, total Abu Dhabi $\Sigma Q'/\Sigma R' = 512/29{,}500 = 1.7\%$

Unweighted mean, large fields: $\Sigma Q/\Sigma R = 394/4{,}959 = 8.0\%$

Weighted mean: $\Sigma[Q(Q/R)]/\Sigma Q = 42.6/394 = 10.8\%$

Sources: Unweighted mean, *OGJ*; others, *IPE*.

escape from some of the ill effects of poorly estimated reserves. Since it is a better predictor of future production, it yields more accurate cost data, which depend on our estimated production profile (Figure 5). We have, of course, a minor sampling problem since we are using the average of the listed fields for the whole country. However, the finite population multiplier is a powerful ally; since we have accounted for 77 percent of the national total of production (394/512), the error cannot be great.

It is likely that outside the United States and Canada, reserves are overstated and therefore decline rates understated. This is not because of errors of optimism, but because undeveloped reservoirs in known fields tend to be counted in.

Table 8 shows the results of this simple approximation, using the weighted average depletion rate as the decline rate. We have chosen four widely differing areas to show the meaning of the concepts, and to test two capacity forecasts. The first case is Mexico, which draws on the data used earlier. It is assumed that Antarctica contains 20 billion bbl available but will not be explored or developed. Hence it has, and will have, zero reserves. Abu Dhabi drilling effort, continued at current levels, leaves them with a 1985 production about where they are now. Iraq attains 4.5 million bbl daily, double current production, but less than the government's announced objective for the year 1980 or 1981.[14]

Optimal Decline Rate. The value of a observed in historical data, and calculated by equation (9), is the result of a particular set of past conditions of

Table 8. Reserves and production of crude oil for selected countries, 1975–1985 (billions of barrels)

Country	*1*	*2*	*3*	*4*	*5*	*6*	*7*	*8*
		Current reserves			*Ultimate**	*Additional proved reserves 1976–1985*	*Production forecast*	
	Cumulative production	*Existing fields*		*Probable*			*1976–1985 Total*	*1985*
		Proved	*Indicated Additional*					
Mexico (Reforma Only)	0.1	5.7	N.A.	N.A.	60.0	19.0	6.4	0.84
Antarctica	0.0	0.0	0.0	0.0	20.0	0.0	0.0	0.0
Abu Dhabi	2.7	5.8	24.2	incl. (3)	49.0	11.0	5.5	0.55
Iraq	10.9	27.8	7.2	incl. (3)	70.0	15.0	12.0	1.64

*Columns 1 + 2 + 3 + 4 plus new fields.

Sources: Col. 1, *IPE*

Col. 2, *IPE*, large fields reserves, multiplied by ratio of their production to total national production, as in Table 5.

Col. 3, *OGJ*, estimate of "proved reserves" *less* column 2.

Col. 5, Private correspondence, and John Moody and R. W. Esser, "An Estimate of the World's Recoverable Crude Oil Resources," paper delivered to World Petroleum Congress, Tokyo, 1975; for Mexico, see Table 6.

Col. 6–8 calculated by methods described in text.

cost, price, and tax. It may or may not be an appropriate guide to future behavior under different conditions, and therefore we should like to be able to calculate this parameter. We can do this by assuming profit-maximizing behavior on the part of oil operators, and solving for the optimal depletion rate, a*, based on estimates of cost per barrel and future price.

First, it is assumed that the profile of capital expenditures E_t shown in Figure 5 can be collapsed to a single-period outlay I. Data on operating costs are rather sketchy; fortunately the bulk of costs are usually capital outlays, and they can be approximated by an adjustment to I, as shown below. Also, to begin, we assume that the capital coefficient, I/Q_p is a constant; later we consider the effects of a coefficient which is higher at higher depletion rates.

By our simplified model of depletion, annual production is $Q_t = Q_p e^{-at}$, and given a constant expected future oil price, P, and discount rate, r, the net present value of a block of reserves becomes

$$\begin{aligned} NPV &= PQ_p \int_0^T e^{-(a+r)t} dt - I \\ &= \frac{PQ_p}{a+r} - I. \end{aligned} \tag{11}$$

By setting $\partial NPV/\partial Q_p = 0$, we can derive the first-order conditions for maximizing the value of the reserves, and the result can be stated in terms of an optimal rate of depletion, a*. That is,

$$a^* = \sqrt{\frac{Q_p}{I} \cdot Pr} - r. \tag{12}$$

Thus the optimal depletion rate is a function of the capital coefficient (or, more properly, its reciprocal), the oil price and the discount rate.

Before proceeding to estimate values of I/Q_p and a*, it is worthwhile to question the accuracy of the simplified or "collapsed" model of oil exploitation in Figure 5. We have reduced a complex process to only three parameters—total proved reserves, total investment and peak output—and we need to make sure that this model yields a reasonable approximation to reality in view of the wide variations in production profiles among fields, and the dependence of the results on the discount rate assumed.

We can get a rough check on this issue by comparing the average per barrel cost of oil as calculated by our "collapsed" model against actual data on expenditure and production profiles for the North Sea. Referring to Figure 5, we can define an oil price, C, which would meet the condition

$$NPV = \int_0^T E_t e^{-rt} dt - \int_0^T CQ_t e^{-rt} dt = 0. \tag{13}$$

This C is the supply price of oil from the reserves illustrated in Figure 5. Or, more usefully for our purposes, C may be referred to as the average cost per barrel of oil from a new project in the area shown. We are able to calculate the "true" value of average per-barrel cost from detailed data for 17 North Sea fields. Similarly, by setting NPV = 0 in equation 11, inserting equivalent values of the "collapsed" parameters, and solving for P, we can calculate a comparable set of figures for the simplified model. When we estimate $C = \alpha_0 + \alpha_1 P$ using the values of C and P thus calculated, the results are as follows:

Discount Rate, r	α_0	α_1	R^2	SE	CV
10	– .028 (–0.19)	0.972 (16.5)	.95	.177	.078
12	– .033 (– 0.18)	0.989 (14.4)	.93	.224	.089
15	– .046 (– 0.18)	1.013 (12.0)	.91	.310	.107
20	– .055 (– 0.13)	1.067 (9.5)	.86	.487	.133

(Numbers in parentheses are t-statistics, SE = standard error of regression, CV = coefficient of variation. Source of data: Wood, MacKenzie and Company, Section 2 of the North Sea Service, February, 1976).

For example, with a 12 percent discount rate, the "collapsed" unit cost is a little higher on the average than the true cost: to get from "collapsed" to true, one multiplies collapsed by 0.989 and subtracts 3.3 cents. Thus estimated, the calculated cost will be within 9 percent of the true cost in about two-thirds of all cases. Whatever the discount rate, however, the collapsed cost is a reasonable approximation to the true cost—though this good fit depends critically on the strict definition of reserves as "planned cumulative output from planned facilities," as discussed above.

Analysis assuming a constant capital coefficient. Unfortunately, outside the North Sea, we do not have anything remotely resembling these detailed investment data. However, capital coefficients for differing producing areas can be estimated very approximately from historical data. Equation (14) shows the method used most frequently to calculate I. For any country, i, we take the proportion of its rig time RY_i to total regional rig time RY. We then multiply this by the Chase Manhattan Bank estimates of total regional production capital expenditures in the year, CMB. That is,

$$I_i = \left[\frac{RY_i}{RY} \cdot CMB \right]. \tag{14}$$

This method overstates oil development costs, since it includes the cost of all wells, including exploratory and gas wells. In terms of Table 5, the cost of adding to total reserves (column 1) is included in the cost of adding to proved reserves (column 4).

In addition to this bias, there is also year-to-year inaccuracy caused by the lumpy nature of much development investment, especially in loading facilities. The investment needed per unit of additional daily output will be exaggerated in a year with much work in process, and conversely understated in the year facilities are completed, and begin operating with little new money being spent. More important: the method overstates costs in any area with an above-average percent of onshore drilling and understates them in an area with an above-average percent offshore, as compared with the whole region. Where this is a problem, corrections can be made using American data.

Additions to capacity Q_p may be calculated in one of two ways (or both as a mutual check). The first is to multiply the number of newly completed wells by the average flow rate of all existing wells.[15] To the maximum extent possible one should divide the completions by separate fields. This method tends to understate capacity increments because new wells are always drilled with better knowledge of the reservoirs than old wells, and to overstate increments because one would expect lower well productivities as well interference began to be significant and as lower-quality strata and reservoirs were developed.

The alternative is to take the difference in production between year t and (t-t′) and add to it the estimated loss of productive capacity over the interim. This involves the calculation of an appropriate decline rate, to be discussed below. Wherever there is irregular fluctuation in output, and any appreciable amount of excess capacity, however, the method cannot be used at all, since change in output is no indication of change in capacity. In this paper, therefore, we used the method of multiplying new completions by average flow rate. However, since independent estimates of capacity are available for recent years, for some countries, we will be able to apply the second method to them.

Table 9 shows capital coefficients for selected countries for the year 1973. (Data for 1974, the only later year, are extremely "noisy.") Since we have used only single-year data, the estimates are subject to such large error factors that they should not be used for any purpose except to check on the calculation of a*, which we now propose to do.

As an example, take a North Sea field where investment per peak daily barrel is calculated to be $7,300. We will take r as 10 percent.[16] The oil price is assumed to be $12.50 per barrel, and initial operating cost (at peak production) is $.95 per barrel,[17] i.e., $347 per year over the life of the field. We can equate this to a capital sum.

Table 9. Capital coefficients: dollars per barrel of initial capacity, selected countries, 1973.

Country	*Capital Coefficient, I/Q_p*
Algeria	$ 758
Libya	144
Egypt	483
Iran	186
Iraq	147
Nigeria	210
Venezuela	1,021
Brazil	5,622
Australia	5,185
Indonesia	335

Sources: Expenditures from Chase Manhattan Bank; Rig-time from Hughes Tool Company and AAPG Bulletin; other data from *OGJ*, "Worldwide Oil," annual issue, 1972–1974, and *World Oil*, "International Outlook," annual issue, 1974.

Assuming the typical North Sea twenty-year life, and a 10 percent discount rate, a stream of outlays or receipts of $347 dollars per year is worth $347/.118 = $2,950. Investment plus operating cost now equates to a capital sum of $7,268 + $2,950 = $10,218. Then

$$a^* = \sqrt{\frac{365(\$12.50)(.10)}{\$10{,}218}} - .10 = .111 .$$

In fact, outside of Auk and Argyll, North Sea Q_p/R ratios (which are an approximation of a*) are about 9 percent. We will see shortly why they should be expected to be a bit lower than optimal.

As another example, we may take the hypothetical field "discovered" by Conoco on the Georges Bank (*Oil and Gas Journal*, July 19, 1976, p. 60). It contains 200 million bbl, to be drained in fifteen–twenty years (say, 17.5 years) and requires an investment of $397 million. We assume operating costs will be slightly lower than Montrose and Heather (slightly smaller fields) in the North Sea, i.e., $25 million annually, over a 17.5-year period, hence with a present value of $203 million, for a total investment of $600 million. Initial output is 60,000 bbl per day, hence the total capital coefficient is $10,000 per daily barrel. Conoco assumes P = $14 per barrel, and hence a* = 12.6 percent. Conoco's estimate is 10.95 percent.

Another comparison is possible, using a paper prepared at the U.S. Geological Survey (12). The authors hypothesize a reservoir with a given quantity of producible gas, reservoir pressure, price-cost parameters, and four possible rates of water encroachment. The operator's decision on number of wells determines the initial Q_p, the ratio Q_p/R, the production profile over time, and the NPV of the deposit. Since price, operating costs, taxes, and the capital coefficient can be approximated from their data, and the interest rate is given, we can calculate $a^* = .135$. In fact, the ratio which would maximize NPV is in the range between .157 and .173. (Although ultimate recovery is sensitive to the rate of withdrawal, there is no well interference, and initial withdrawal is explicitly stated as strictly proportional to the number of wells drilled.)

Variable Capital Coefficients. So far, it would appear that a* gives values not too far out of line with reality. But the three validation examples have been very high-cost areas. Typical capital coefficients are only a small fraction of such values. Yet observed values of the depletion rate are not many times those calculated; in fact, they are typically lower.

Many of the low observed rates are apparently not real, because of overstated reserves, as shown above. But we might consider the United States, where reserves are estimates strictly on a very large sample of reservoirs. Since 1972 when the United States essentially went on maximum efficient rate (MER) capacity production, the ratio of production to proved reserves has been very stable in the neighborhood of 11.8 percent. It is expected to remain there, and production out of proved reserves to decline at that rate, (14), p. 214.

The incentive to speed up recovery has greatly increased in 1972–1975, but the rate has stayed constant. Here we might reckon that a worldwide average limit to the depletion rate out of proved reserves is approximately 12 percent, and in few fields will it exceed 20 percent. One obvious reason is that too high a rate would induce gas or water breakthrough to the well bore, leaving much oil behind, reducing reserves.

Moreover, we have so far assumed that the investment per daily barrel remains constant regardless of how intensively the field is developed. In fact, in many cases the value of I/Q_p may not be a constant over wide-ranging values of a. It is in fact constant over some range, wide or narrow, for the whole hydrodynamic system is involved: the oil in place (including dissolved gas), any associated gas, and water. The volume and pressure of the hydrocarbon may be a relatively minor fraction of the whole. But usually, past some point the higher the output, the lower the output per well. Hence, there is a margin where the output gained by drilling a well declines, perhaps sharply. Therefore, we may need to make I, the investment, an increasing function of Q_p/R. To do this, James L. Smith has devised the following method.

We specify the reservoir investment function as

$$I = Ka^{\varepsilon}R \tag{15}$$

with K the proportionality constant and ε (epsilon) the elasticity of investment with respect to the depletion rate. Since $aR = Q_p$, a varying ε in $a^{\varepsilon}R$ will vary the investment as a function of chosen output (i.e., the depletion rate). The constant K merely translates planned initial output into dollar requirements. Thus $\varepsilon = 1$ implies constant (linear) costs; $\varepsilon > 1$ implies increasing costs, etc.

The equivalent of Equation 11 is

$$NPV = \frac{PQ_p}{a + r} - Ka^{\varepsilon}R, \tag{11a}$$

and the optimal decline rate can be expressed in the following relation

$$a^* = \sqrt{\frac{Pr}{\varepsilon K(a^*)^{\varepsilon - 1}}} - r. \tag{12a}$$

Note that equation (12) is just a special case of 12a, where $\varepsilon = 1$ and therefore $K = I/Q_p$. The a* in equation (12) implied that well productivity had not yet entered the phase of diminishing returns. But equation (12a) now lets us obtain a marginal cost which allows for the increasing costs imposed by well interference, and to derive an optimal initial producing rate accordingly.

With an estimate of ε, we have all the parameters needed to solve for a*, with the exception of the constant K. By equation (15) we can state K as a function of I/Q_p and a, and we have the historical value of the capital coefficient, $\hat{I}/\hat{Q}_p$, and the depletion rate $\hat{a}$, which are proxies for the true values. Thus

$$K = [\hat{I}/\hat{Q}_p]\hat{a}^{1-\varepsilon}. \tag{15a}$$

With K as estimated by 15a, we can solve equation (12a) (we use an iterative algorithm that permits a numerical approximation) for the optimal *future* depletion rate a*.

Table 10 shows some of the results, assuming the historical depletion rate was 0.1, and for P = \$10.50 and r = .12. The reason why we need this

Table 10. Optional depletion rate (a*), for $\hat{a} = 0.1$, P = \$10.50, r = .12.

Elasticity	*Historical capital coefficient* ($\hat{I}/\hat{Q}_p$)			
(ε)	\$300	\$800	\$1,500	\$8,000
1.0	1.12	0.64	0.43	0.12
1.25	0.74	0.44	0.31	0.10
1.5	0.54	0.34	0.24	0.08
2.0	0.35	0.23	0.18	0.08
3.0	0.21	0.16	0.13	0.07

procedure is not far to seek. The range of observed capital coefficients is enormous. The top line, which puts the interference effect to zero (i.e., $\varepsilon = 1$), would indicate that with particularly low capital requirements ($300 per initial daily barrel) the oil should be brought out in less than a year. If the operator of the reservoir acts as a competitive price-taker, then he may be restrained by the loss of ultimate recovery, or by well interference expressed in the ε coefficient. If the operator is not inhibited, and yet is producing at "too low" a depletion rate, then there is slack in the system which we would expect him to use up, by expanding output. If none of these inhibitors is sufficient, then he may be restrained by fear of spoiling the market.

The decline in per-well productivity is a prominent feature of many field production histories. But these are all after the fact. The basic data that would permit a calculation of an ex-ante a* are rarely, if ever, published. Filling this gap is an obviously high-priority task.

A final qualification: an expectation of higher prices would *ceteris paribus* justify delaying output. This is a problem in optimal control theory, and we intend to work out the functional relation.

The Influence of Taxes. Taxation may lower the optimal output rate. Looking again at the North Sea example and assuming output to have been pushed to where $\varepsilon = 1.0$, NPV is maximized by depleting at 11 percent per annum. From equation (11) and the data on p. 54, we can calculate the DCF rate of return r′. That is, if NPV = 0 in Equation 14, then

$$\frac{PQ_p}{a + r'} = I, \tag{16}$$

and in this case $4,562/(.11 + r′) = $10,218 and thus r′ = 0.34.

The objective of the government is to capture as much as possible of the difference between a DCF of 34 percent, and the operator's minimum acceptable DCF of, say, 10 percent, which is how we define the bare cost of production. But royalty or excise payments may make the operator change his plans, to make everybody worse off. For example, if they simply took a 50 percent royalty, thus cutting P to $6.25, the operator would maximize NPV at a* = .05.

Thus the investment and the peak output would be only 5/11 of optimal. It is not worth the operator's while to spend more money to get the oil out faster, though he will get it out eventually.[18] The total NPV per barrel is less:

$$NPV/Q = ((5/11)(365 \times \$12.50)/.15) - \$10{,}218 = \$3{,}608.$$

This is only 45 percent of the optimal NPV.

$$\text{The operator's DCF is now: } \$10{,}218 + \frac{(\$6.25(365)5/11)}{.05 + r'} = 0$$

$$r' = .10 - .05 = .05,$$

which is the course unacceptably low, and the government must prepare a lower royalty. Where costs are very low, the dampening effect of a royalty or excise does not matter nearly as much. Given a not infrequent case of poor information and mutual mistrust, one can easily write such a scenario of deadlock as is being played out in several countries today. An alternative would be to set NPV as the upper limit to taxation, and aim to get some maximum practical fraction of it. The chief barrier is uncertainty. Where there is agreement on the cost and revenue data in our example, the government could bargain for a lump sum payment of somewhat less than $11,500, payable in installments at the convenience of both parties, but independent of the volume of production, so that the operator would have no incentive to change his plan. Another possibility would be to calculate total NPV as expected, then to provide a sliding scale of government take, in order to give the operator an incentive to reduce costs or increase output.

We may eventually devise some plan, but for the present we need only note that since governments have not followed it, they must have reduced the rate of production and NPV below what is attainable. Again, we follow a simplified approach: translate the tax into a royalty-equivalent, of the type of equation (16), and calculate with the resulting depletion rate. In practice this becomes very complex, because the usual arrangement is for recovery of costs at an early stage. This reduces, often drastically, the present-value equivalent excise, and, therefore, the impact on NPV.

CONCLUSION

Oil costs are everywhere only a small fraction of prevailing prices. Hence even substantial price changes would have little effect on supply. Moreover, the owners of the resource are governments, with more than the usual number of degrees of freedom to choose investment and pricing policies. Hence a model driven by some assumed price-cost-profit equilibrium will probably not capture the essentials of the supply side of the market. Furthermore the basic determinants of cost and supply—the investment needed to find, delineate, and exploit reservoirs, and the time for this operation—are so imperfectly known that the need to respect data limitations has dominated our model. We have perforce adopted a rough and often ad hoc scheme for predicting the amount of reserves to be developed, calculating real production costs, and predicting government policy on capturing the profits on incremental investment, and deciding on the rate of growth of capacity.

For most countries, with relatively small reserves, a rather mechanical procedure and assumed zero supply elasticity results in no substantial error. For those relatively few countries that matter, we hope to supplant our first working assumptions with more realistic and complex hypotheses, including even backward bending supply responses.

FOOTNOTES

This paper represents a collective effort by the Supply Analysis Group of the MIT World Oil Project. Included in the group are Professors Gordon Kaufman (MIT) and Eytan Barouch (Clarkson College); Dr. Paul L. Eckbo (Norwegian School of Economics and Business Administration); and research assistants James Paddock, James Smith, and Arlie Sterling. An earlier version was given at the International Energy Agency's Workshop on Energy Supply, November 1976.

The Project is supported by the U.S. National Science Foundation under Grant No. SIA75–00739. However, any opinions, findings, conclusions or recommendations expressed herein are those of the authors and do not necessarily reflect the views of NSF.

1. An overview of the research method, and the results of the early simulation studies, are shown in the work of Eckbo (8).

2. Examples of this type of model include those by Pindyck (18), Hnyilicza and Pindyck (10), and Cremer and Weitzman (7). Price scenarios based on judgment or the analysis of others also can be tested using the simulation framework.

3. For an example, see the work of MacAvoy and Pindyck (11).

4. What follows is an abbreviated version of the disaggregated method. Its development and initial applications are described in a paper by the MIT World Oil Project (19), and the methodology is further developed in a paper by Eckbo, *et al.* (9).

5. The estimates are treated as measures of "proved" reserves, although the strict conditions for such a definition often have not been met (i.e., limiting the estimate to reserves actually enclosed by development wells). The assumption is that a small number of exploratory and development wells, when coupled with geophysical data on structure size, gives a reasonably accurate estimate of what will ultimately be proved in a given specific reservoir.

6. A similar phenomenon is built into the aggregated analysis as well, where the amount added to reserves is a declining function of the number of rig years in an area.

7. Note that the discovery process has been defined in terms of recoverable reserves, and the analysis of equations (1) through (4) carried out apart from concern for the oil price. In principle, the amount of oil that is "recoverable" from the oil in place is a function of price and cost factors: at higher oil prices it is worthwhile to spend more to recover a larger percentage of the oil. In practice, given the state of the art at any moment, and in the relevant range of prices, elasticity of recovery factor to price appears to be very low.

8. In the Soviet Union, which operates nearly as many rigs as the United States, 25–40 percent of tne time is used in moving them. See *World Oil*, August 15, 1976, p. 126.

9. *Petroleo Y Otros Datos Estadisticos, 1974* (Republic of Venezuela, Ministry of Mines and Hydrocarbons), Table, page 45. Although revisions were 5.5 billion, nearly 1 billion barrels must have been contributed outside the "markup."

10. See (1), Table III, respective years.

11. For a fixed Ult R_t (i.e., no major change in the ultimate prospect) the "discovery decline" is linear. As Cum R_t goes from Cum R_{75} to Ult R_t, b_t goes (linearly) from one to zero.

12. The Mexican authorities use a definition of proved reserves which is very close

to the strict API concept: "reserves which are expected to be produced by existing wells through primary and secondary recovery"; from *Prospectus, Mexico External Bonds Due 1983* (September 1976), p. 17.

13. Of course, we need to worry about the degree of error this simplified model may involve. The North Sea affords a check. The end-1973 *IPE* reserves (fields under development only) were less than 2 billion bbl. Preliminary indications then (which have, incidentally, been well borne out) were of a 9 percent depletion rate. This would predict 495 tbd in 1977, somewhat over a year too late, since 1976 and 1977 are now estimated at 585 and 1,230 tbd, respectively. The end-1975 *IPE* developed reserves were 9.4 billion bbl, and the same method, again assuming 9 percent, would predict 2.3 mbd in 1979, which seems about on target so far, since late-1976 estimates are for 2.5 mbd in 1979.

14. The equating of capacity with production assumes that these cartel nations are not forced to hold excess capacity in order to help support the price.

15. Our estimate of the weighted average daily flow rate per well for each country is calculated as follows:

$$\text{estimated flow rate} = \sum_{i=1}^{M} \left(\frac{Q_i}{Q_M}\right)\left(\frac{Q_i}{W_i}\right),$$

where M is the number of major fields (a major field is defined as one which accounts for at least 3 percent of the country's total production), Q_i is the production of field i, Q_M is the total production of all major fields, and W_i is the number of producing wells in field i. Thus

$$\frac{Q_i}{Q_M} \text{ is our weighting factor and } \sum_{i=1}^{M} \frac{Q_i}{Q_M} = 1.$$

16. During 1975–1976, dollar denominated Eurobonds had a 9 percent yield, which we consider as the prime business-risk rate on long-term finance. We add 1 percent, which is the usual premium above the LIBOR (London Inter-Bank Rate) on secured financing for the North Sea development projects.

Perhaps a better inflation allowance would start from the fact that the sterling LIBOR was generally 4 percentage points above the dollar LIBOR in 1975–1976 (9.5 and 5.5 percent, respectively) before the sterling slide in late 1976. An interesting confirmation of our 9 percent discount rate is in the LASMO-SCOT "package" rate, in late 1976, which can be shown to approximate 14 percent in sterling (see the LASMO-SCOT *Prospectus*, February 1976).

17. Wood-MacKenzie report, October 1976, Section Two, Table 1, average for all proved fields.

18. Note that one barrel per day, declining at 10 percent, cumulates to 9.09 barrels of original reserves. If we produce only (5/11) of a barrel, declining at 5 percent, this too cumulates to 9.09 barrels.

REFERENCES

1. American Petroleum Institute, American Gas Association, and Canadian Petroleum Institute, *Reserves of Crude Oil, Natural Gas Liquids, and Natural Gas in the United States and Canada as of December 31, 1975*, Vol. 30, May 1976.
2. Barouch, E., and G. M. Kaufman, "Oil and Gas Discovery Modelled as Random Proportional to Random Size," MIT Sloan School of Management, Working Paper WP 888–76, December 1976.

3. ———, "Probabilistic Modelling of Oil and Gas Discovery," in *Energy: Mathematics and Models*, SIAM Institute for Mathematics and Society, 1976.
4. Beall, A. O., "Dynamics of Petroleum Industry Investment in the North Sea," MIT Energy Laboratory Working Paper No. 76–007, June 1976.
5. J. Birks, "The North Sea in 1973," North and Celtic Seas Conference, London, 1973.
6. Crandall M. S., "The Economics of Iranian Oil," MIT Energy Laboratory Working Paper No. 75–003, March 1975.
7. Cremer J., and M. L. Weitzman, "OPEC and the Monopoly Price of World Oil," *European Economic Review*, August 1976.
8. Eckbo P. L., *The Future of World Oil*, Ballinger Publishing Company, Cambridge, Mass., 1976.
9. ———, H. D. Jacoby, and J. L. Smith, "Oil Supply Forecasting: A Disaggregated Process Approach," *Bell Journal of Economics*, Spring 1978.
10. Hnyilicaza E., and R. S. Pindyck, "Pricing Policies for a Two-Part Exhaustible Resource Cartel: The Case of OPEC," *European Economic Review*, August 1976.
11. MacAvoy P. W., and R. S. Pindyck, *The Economics of the Natural Gas Shortage (1960–1980)*, North Holland, Amsterdam, 1975.
12. McFarland, J. W., T. E. Springer, E. A. Monash, and R. G. Cummings, "Long Term Alternatives Between Production Rate and Supply of Oil and Natural Gas: Some Implications From a Simulation Study" (n.d.).
13. MIT World Oil Project, "Progress on Analysis of the World Oil Market," MIT Energy Laboratory Working Paper No. 75–015, October 1975.
14. National Petroleum Council, *Enhanced Oil Recovery*, December 1976.
15. National Petroleum Council, *U.S. Energy Outlook*, December 1972.
16. Odell, P. R., and K. F. Rosing, "The North Sea Province: A Simulation Model of Development," *Energy Policy*, December 1974.
17. Organization for Economic Cooperation and Development, *Energy Prospects to 1985*, 1974.
18. Pindyck, R. S., "Gains to Producers from the Cartelization of Exhaustible Resources," MIT Energy Laboratory Working Paper No. 76–012, May 1976.
19. Supply Working Group, MIT World Oil Project, "Supply Forecasting Using Disaggregated Pool Analysis," MIT Energy Laboratory Working Paper No. 76–009, May 17, 1976.
20. U.S. Federal Energy Administration, *Project Independence Report*, U.S. Government Printing Office, Washington, D.C., November 1974.

A BASIN DEVELOPMENT MODEL OF OIL SUPPLY

Paul Leo Eckbo, MASSACHUSETTS INSTITUTE OF TECHNOLOGY AND THE NORWEGIAN SCHOOL OF ECONOMICS AND BUSINESS ADMINISTRATION

Some serious shortcomings of the traditional approach to supply analysis become apparent when applied to extractive industries. A framework for analysis of an extractive industry should include a representation of the geological resource base and how this resource base changes as a result of exploration and production activities. The tax regime should be included as taxes in extractive industries constitute a larger and more distortive element of cost than in conventional industries. However, traditional methods most often approach the problem of supply estimation by formulating and estimating the parameters of the production function or its counterpart, the cost function. Neither the resource base nor the tax regime is

Advances in the Economics of Energy and Resources, Vol. 2, pp. 39–58.
Copyright © 1979 by JAI Press, Inc.
All rights of reproduction in any form reserved.
ISBN: 0-89232-079-6

adequately reflected when supply analysis is confined to conventional concepts of the production function.

Due to the difficulties surrounding specification and aggregation of production functions for individual plants, most researchers attempt to estimate directly the production function or the cost function which represents the industry as a whole. This approach is easily defensible when applied to industries where producing units are uniform and the micro-production relationships are stable through time. However, in the oil industry the need to carry out the supply analysis on a disaggregated, "plant" level is accentuated by the heterogeneous nature of the stock of resource deposits. The target of exploration and the unit of production (the "plant") in the oil industry is the reservoir. The reservoir is thus the natural unit of analysis when analyzing the supply of oil.

Supply activities fall into two distinctly different categories: the exploration for and discovery of reservoirs, and the development and production of those reservoirs found. From an analytical point of view a framework designed to represent the exploration activities is different from a framework designed to represent the development and production activities. The link between the two stages of production is obvious. The economic incentive to explore is the net present value of producing the reservoir that might be found. At the margin, the cost of developing and producing an additional barrel of oil from an existing reservoir should be equal to the certainty equivalent cost of extracting an additional barrel of oil from a reservoir to be found. By separating exploration and finding activities from development and production activities it is possible to capture the two major aspects of resource depletion, the depletion of producible reservoirs from the population of reservoirs to be found and the depletion of recoverable reserves from the existing population of producible reservoirs.

The separation of the discovery/development process is an attempt to separate the geological characteristics of an area from the economic attractiveness of the area. This separation, however, is not complete. The selection of the prospects to be drilled by the explorer is influenced by the expected economics of the reservoirs to be discovered. However, this separation does allow the risk involved in exploring for hydrocarbons to be subdivided into two categories: the geologic risk and the economic risk. The geologic risk results from the lack of a technology to determine, prior to drilling, whether a prospect[1] is, in fact, a reservoir, i.e., contains hydrocarbons. This geologic risk is represented by the probability of hitting a prospect with no hydrocarbon fluids. The economic risk results from the fact that some of the reservoirs discovered may not contain hydrocarbons in commercially attractive quantities.

As part of a study of the world oil market (6, 15), we are exploring new approaches to the analysis of oil supply (1,2,3,8,14). The most ambitious of these undertakings is an approach wherein the analyst specifies and (where possible) estimates structural relationships which relate factor inputs and outputs at each stage of the supply process: exploration, discovery, reservoir development, and production. Then, contingent on prices, factor costs, the tax regime, and miscellaneous public policy constraints, activity levels are determined which optimize the economic return to petroleum operators (8,14). This approach is labeled a disaggregated process approach and is also described in the Adelman and Jacoby paper (1). An integral part of that approach is a statistical analysis of the exploratory process on the play[2] level (2,3,11,12). This statistical analysis is anchored in an explicit hypothesis about the nature of geological deposition, and a related assumption about the characteristics of the exploratory process. With these hypotheses, it is possible to predict the density functions of future discoveries conditional upon the exploratory history already experienced.

The focus of the Basin Model is on the supply potential of a basin.[3] A basin will, in general, consist of a number of plays. However, the statistical analysis on the play level is not easily aggregated to the basin level. No satisfactory statistical representation exists of the process by which plays are sought and discovered. Rather than relying on a statistical analysis of the exploration process, the Basin Model, therefore, relies on a simple exponential decline function to generate a discovery sequence from an exogenous rate of exploratory drilling. The Basin Model is thus a disaggregated process model (1, 8, 14) in which the statistical analysis of the exploration process has been replaced by a deterministic discovery decline relationship. The reservoirs generated by both approaches to discovery analysis enter into the same reservoir model which takes account of costs and expected future petroleum prices, and allows detailed consideration of the tax regime.

The cost relationships and the development and production profile of the reservoir model replaces the capital coefficient and the production decline relationship of the "aggregated country approach," as discussed in the Adelman and Jacoby paper (1). The capital coefficient is estimated as a historic average of the cost of an additional barrel of oil-producing capacity. Hence, the unit capital cost concept of the aggregated country analysis will not reflect the future level of marginal cost. The cost relationships of the reservoir model are specified and estimated as functions of reservoir size. The discovery decline relationship produces decreasing size of the reservoirs found. By feeding the decreasing reservoir sizes into the reservoir model, the Basin Model will account for the marginal cost of a barrel to be developed and produced. The Basin Model is thus a marginal cost model as opposed to the average cost focus of the aggregated country analysis. The marginal cost focus is made possible by defining the reservoir as the unit of analysis rather than regional additions to recoverable reserves.

The level of ultimate recoverable reserves is specified as an exogenous variable in the aggregated country analysis. This statistic is of vital importance to the understanding of the supply potential of an area. Geological, technological, economic, and policy variables affect the ultimate level of recoverable reserves. The assumptions made with respect to each of these variables are seldom made explicit. The ultimate level of recoverable reserves is endogenous to the Basin Model. The Basin Model is thus a device to disentangle the assumptions embedded in estimates of ultimate recoverable reserves.

The statistical approach to discovery analysis requires both a detailed documentation of the discovery history of the area and an extensive computational effort. Such a detailed documentation of the discovery history is rarely available in the public domain and for many potentially interesting areas there is no discovery history. The Basin Model was developed to require much less information on the geology and discovery history of an area. The much greater uncertainty involved in analyzing areas with limited or no discovery history makes it reasonable to trade off some conceptual richness and complexity for reduced estimation and simulation costs. The Basin Model allows scenario-type simulation to be undertaken at a low cost. When the uncertainty surrounding the underlying key parameter values is considerable and not easily quantifiable, sensitivity analysis simulations are a useful way of learning about the supply potential of an area.

Section 1 is devoted to a discussion of a key aspect of the problem of supply forecasting, the discovery decline phenomenon. In section 2, the Basin Model is presented and applied to the North Sea. In section 3, an attempt is made to validate the model. The elements of the elasticity of oil supply are discussed in section 4. Section 5 concludes this discussion of the Basin Development Model and indicates some areas of future research.

1. THE DISCOVERY DECLINE PHENOMENON

The disaggregated process model draws on research by Kaufman and Barouch (2, 3) on statistical analysis of the exploratory process. The analysis is based on two hypotheses, one about the nature of resource deposition and the other on the character of oil exploration. It is assumed that reservoir size, measured in terms of recoverable reserves, r, is a random variable, and that its density function is lognormal (2, 3, 10, 11). To this hypothesis about nature is added an hypothesis about the process by which oil operators search for and find reservoirs. Following the work of Kaufman and others, the exploratory process is characterized as one of random sampling, without replacement, in proportion to reservoir size r. With these hypotheses, it is possible to predict the density functions of future discoveries conditional upon the exploratory history already experienced. In a play in which a total of N discoveries are to be made as the play is "drilled up," the sequence of remaining discoveries (r_j, $j = n, \ldots, N$) is completed described by a sequence of predictive density functions:

$$P(x_1 < r_j \leq x_2) = \int_{x_1}^{x_2} f_j(r|\hat{\mu}, \hat{\sigma}_j^2 \hat{N})dr \qquad j = n, \ldots N, \tag{1}$$

where $P(x_1 < r_j \leq x_2)$ is the probability that a discovery is within a size class defined by x_1 and x_2, and $\hat{\mu}, \hat{\sigma}$ are the sample mean and variance of log r_j and $\hat{N}$ the expected value of N.[4]

The risk involved in the exploratory process may be split into two components: the geologic risk or probability of a dry well, and the economic risk or probability that the reservoir found is too small to be economically produced. We may consider the economic risk associated with discovery j by integrating equation (1) over the range of economically viable reservoir sizes when the expected contribution to the recoverable reserve base from discovery j, ER_j, is calculated. The minimum economic reservoir size, MERS, defines the borderline of economic viability with respect to reservoir size. The expected contribution to the recoverable reserves base from discovery number j, E (ER_j) is therefore:

$$E(ER_j) = \int_{MERS}^{\infty} r \cdot f_j(r|\hat{\mu}, \hat{\sigma}_j^2 \hat{N})dr \qquad j = n \ldots N \tag{2}$$

By integrating equation (2) over reservoir sizes larger than the MERS, the variability of actual reservoir size around the predictive mean is recognized. Equation (2) is a special case of the general problem of estimating "partial expectations," PE, for a discovery sequence, i.e., the expected number of barrels to be found in reservoirs of various sizes. By defining a set of size classes $\{S_i\}$, the expression for the expected number of barrels to be found in reservoirs of size class i for discovery j, the partial expectation of size class i for discovery j, then becomes

$$PE(r_j|S_i) = \int_{r \varepsilon S_i} r \cdot f_j(r|\hat{\mu}, \hat{\sigma}_j^2 \hat{N})dr \qquad j = n, \ldots, N. \tag{3}$$

We may define the expected number of discoveries in period t, d_t, to be the product of the number of exploratory wells drilled in period t, w_t, and the probability of identifying a reservoir, PS,

$$d_t = w_t * PS. \tag{4}$$

The cumulative number of discoveries made prior to period t, cum_t, is defined as follows:

$$cum_t = \sum_{t=1}^{t-1} d_t. \tag{5}$$

$$cum_1 = 0.$$

The total increment to recoverable reserves in year t, RA_t, consists of all current discoveries of oil in reservoirs which satisfy the MERS criterion, plus any oil from earlier discoveries which for the first time also satisfy this criterion. The cumulative inventory of submarginal reservoirs is thus monitored continuously as economic incentives fluctuate. If we label the level MERS in period t, M_t, then the expression for the total annual increment to recoverable reserves is:

$$RA_t = \sum_{S_i > M_t} \sum_{j=1}^{d_t} PE(r_j|S_i, cum_t) + \sum_{M_{t-1} > S_i \geq M_t} \sum_{j=1}^{cum_t} PR(r_j|S_i). \tag{6}$$

the first part of equation (6) summarizes the contribution to recoverable reserves from current discoveries. The second part of the equation summarizes the contribution of already discovered but previously submarginal reservoirs. The second part of equation (6) is written in an "irreversible" form, i.e., once a reservoir has entered the development stage it will not be thrown back into the inventory of submarginal reservoirs. A more complex version of equation (6) is required to keep track of the inventory of submarginal reservoirs if MERS is to be allowed to fluctuate rather than only gradually to decrease. Equation (6) illustrates, however, the "inventory control" aspect of discovery analysis, or the fact that submarginal reservoirs found are held in an "inventory" to await a more favorable economic environment.

Using the North Sea as an example, and assuming that the 60 discoveries of Table 1 belong to one play (as opposed to the four plays into which the discoveries might have been divided), the parameters of the lognormal distribution were estimated in addition to the expected total number of reservoirs to be found in the Northern North Sea (2,8,14). A sequence of predictive density functions was estimated, as were the partial expectations associated with the next 92 discoveries. We may summarize the results of this estimation and simulation exercise by fitting some simplified functions to the outcome of the calculation of the expected discovery sequence. A simple exponential decline function was fitted to the sequence of expected discovery sizes $E(r_j)$, as well as to the partial expectations corresponding to five separate size classes (oil and gas equivalents): $S_1 = 0 - 125$ million barrels of recoverable reserves (MMB), $S_2 = 125 - 250$ MMB, $S_3 = 250 - 375$ MMB, $S_4 = 375 - 500$ MMB, and $S_5 = 500 - \infty$ MMB. The partial expectations were estimated as fractions of the expected discovery size.

$$E(r_j) = 526.347\, e^{-0{,}0123j}; \qquad j = 61, \ldots, 152\,(R^2 = 0.99), \tag{7}$$

$$PE(r_j|S_1) = [526.347\, e^{-0{,}0123j}]\left[0.0251 + 0.0008\left(\frac{j-60}{200}\right) - \right.$$
$$\left. - 0.0006\left(\frac{j-60}{200}\right)^2\right] = \left(1 - \sum_{i=2}^{5} PE(r_j|S_i)\right), \tag{8}$$

Table 1. Northern North Sea discoveries, recoverable reserves—oil and gas. Oil Equivalent (Millions of Barrels)

Order (*j*)	*Name or Location*	*Drilling start-up date*	*Size* (r_j)	*Order* (*j*)	*Name or Location*	*Drilling start-up date*	*Size* (r_j)
1	Cod	2/68	156	31	Ninian	9/73	1000
2	Montrose	4/69	180	32	Statfjord	12/73	4960
3	Ekofisk	9/69	1713	33	Odin	12/73	178
4	Josephine	6/70	100	34	Bruce	3/74	450
5	Tor	8/70	243	35	Magnus	4/74	800
6	Eldfisk	8/70	919	36	N. E. Frigg	4/74	71
7	Forties	8/70	1800	37	Balder	4/74	100
8	W. Ekofisk	8/70	490	38	Andrew	4/74	300
9	Auk	9/70	50	39	Claymore	4/74	375
10	Frigg	4/71	1325	40	E. Magnus	6/74	250
11	Brent	5/71	2500	41	9/13–4	6/74	220
12	Argyll	6/71	70	42	15/6–1	9/74	150
13	Bream	12/71	75	43	Brae	9/74	800
14	Lomond	2/72	500	44	Sleipner	11/74	50
15	S. E. Tor	4/72	34	45	Hod	11/74	75
16	Beryl	5/72	525	46	211/27–3	11/74	450
17	Cormorant	6/72	165	47	Gudrun	11/74	450
18	Edda	6/72	98	48	2/10–1	11/74	100
19	Heimdal	7/72	300	49	3/4–4	12/74	100
20	Albuskjell	7/72	357	50	14/20–1	1/75	75
21	Thistle	7/72	375	51	Crawford	1/75	150
22	Piper	11/72	638	52	9/13–7	1/75	350
23	Maureen	11/72	500	53	3/8–3	1/75	100
24	Dunlin	4/73	435	54	Tern	2/75	175
25	3/15–2	4/73	150	55	21/2–1	2/75	175
26	Hutton	7/73	250	56	3/2–1A	3/75	200
27	Alwyn	7/73	350	57	Valhalla	4/75	50
28	E. Frigg	8/73	623	58	3/4–6 & 3/9–1		200
29	Heather	8/73	150	59	15/13–2		200
30	Brisling	8/73	75	60	211/26–4		175

Sources: Beall (4), and estimates by the MIT World Oil Supply Group as of June 1976.

$$\mathrm{PE}(r_j|S_2) = [526.347\, e^{-0{,}0123j}]\left[0.0714 + 0.0146\left(\frac{j-60}{200}\right) + 0.0113\left(\frac{j-60}{200}\right)^2\right] (R^2 = 0.998), \tag{9}$$

$$\mathrm{PE}(r_j|S_3) = [526.347\, e^{-0{,}0123j}]\left[0.1006 + 0.41\left(\frac{j-60}{200}\right) + 0.0406\left(\frac{j-60}{200}\right)^2\right] (R^2 = 0.998), \tag{10}$$

$$PE(r_j|S_4) = [526.347\,e^{-0,0123j}]\left[0.1223 + 0.0717\left(\frac{j-60}{200}\right) + \right.$$

$$\left. + 0.0898\left(\frac{j-60}{200}\right)^2\right](R^2 = 0.998), \tag{11}$$

$$PE(r_j|S_5) + [526.347\,e^{-0,0,123j}]\left[0.6806 - 0.1281\left(\frac{j-60}{200}\right) + \right.$$

$$\left. + 0.1423\left(\frac{j-60}{200}\right)^2\right](R^2 = 0.997). \tag{12}$$

It is apparent from the high R^2's that these simplified functions capture rather well the characteristics of the predicted discovery sequence of the statistical analysis when applied to the North Sea basin as a whole. No R^2 is reported for equation (8) which was estimated as a residual. Equation (8) says that about 2.5 percent of the additions to reserves are expected to be in reservoirs in size category one; however, this fraction will fluctuate as the exploratory history unfolds, as indicated by the quadratic form of the equation. We may simplify the reporting of the outcome of the analysis even further by stating that the North Sea basin analysis indicates that the average expected discovery decline rate in the North Sea is 1.23 percent.

2. THE BASIN MODEL

The general form of the exponential discovery decline function estimated above is:

$$D_j = L \cdot e^{-b \cdot j}. \tag{13}$$

The expected size of discovery number j, D_j, is a function of a constant L that determines the location of the decline curve, the location of discovery j in the sequence of discoveries as indicated by j, and the discovery decline rate b. We incorporated the economics of reservoir development and production in the discovery analysis by introducing the concept of minimum economic reservoir size, MERS. A generalized one equation representation of a calculation of the present value, PV, of a reservoir would be:

$$PV = \sum_{t=1}^{T} \frac{P^t Q^t(r_j) - I^t(r_j) + d^t - F^t(Q^t(r_j)\pi^t) - K^t - O^t(r_j)}{(1+i)^t}. \tag{14}$$

The notation is to be interpreted as follows: T is the time-horizon of development and production of a reservoir, P is the price of oil, Q is annual production as a function of recoverable reserves, I is investment, d is the amount of investment financed by borrowed capital, F is the taxes which are a function of both the rate of production and the rate of operating profits, π^t, K^t represents the servicing of debt in period t, O^t is operating costs, and i is the discount factor.

The minimum economic reservoir size corresponding to price expectations P^t is determined by the following identity:

$$PV \equiv 0$$

$$\equiv \sum_{t=1}^{T} \frac{P^tQ^t(MERS) - I^t(MERS) + d^t - F^t(Q^t(MERS), \pi^t) - K^t - 0^t(MERS)}{(1+i)^t}. \tag{15}$$

The economic incentive to drill is positive as long as the expected size of the next discovery is larger than or equal to the MERS. We may calculate the expected number of economically viable reservoirs in a basin, N, by exploiting the fact that the last discovery to be made is expected to equal the MERS:

$$D_N \equiv MERS$$

$$MERS = Le^{-b \cdot N}$$

$$N = \frac{-\ln\left(\frac{MERS}{L}\right)}{b}. \tag{16}$$

Equation (16) allows us to express the ultimate level of recoverable reserves in an area, UR, as a function of the geology of the area (as summarized by the discovery decline function) and the economics of the area (as summarized by the MERS-relationship):

$$UR = \int_0^N Le^{-b}dj = \left[-\frac{1}{b}Le^{-b,j} \right]_O^N = \frac{L}{b}\left(1 - e^{-bN}\right), \tag{17}$$

or by substituting equation (16) for N:

$$UR = \frac{L}{b}\left(1 - \frac{MERS}{L}\right). \tag{17}$$

By a simple transformation of equation (17) we get an expression for the discovery decline rate:

$$b = \frac{L}{UR}\left(1 - \frac{MERS}{L}\right). \tag{18}$$

From equation (17) we may estimate the price-sensitivity of the level of ultimate recoverable reserves in a simplified way. An increase in the price of oil will decrease MERS which again will increase N and the integral under the discovery decline curve, UR. Diagrammatically the analysis may be portrayed as in Figure 1.

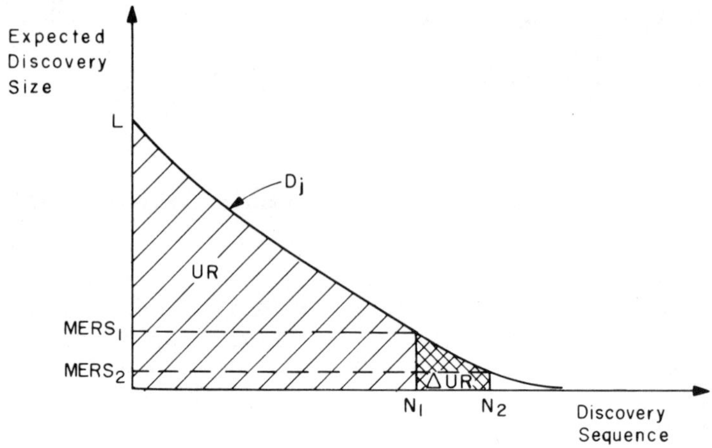

Figure 1. Discovery decline and ultimate recoverable reserves.

By focusing on the expected size of each discovery in a discovery sequence, this discovery analysis does not enable us to consider the "inventory control" problem, i.e., the handling of submarginal reserves discovered in the process due to the variance around the mean discovery value. In the North Sea the MERS is less than 125 MMB, i.e., inside the smallest size category, S_1. A MERS of 125 MMB corresponds to a price of about \$9 (in 1976 dollars) (5,8). Equation (8) indicates that failure to consider the inventory aspect would lead to an overestimation of total North Sea reserves by about 2.5 percent assuming that the discovery decline curve is located as determined by the statistical analysis.

To indicate the locational bias introduced in the discovery decline curve by the exponential model, the exponential decline relationship of equation (13) was estimated directly from the 60 North Sea discoveries of Table 1. The assumption about the lognormality of the reservoir population and the assumption that the exploration process may be represented as a process of sampling without replacement and proportional to reservoir size is then replaced by the simple implicit assumption of the exponential discovery decline function that on the average the largest reservoirs tend to be found first. The exponential decline function was chosen over other decline functions due to the goodness of fit when fitted to the outcome of the statistical analysis of the North Sea.

The estimated exponential discovery decline function for the North Sea emerged as follows (oil and gas equivalents):

$$D_j = 764.017\, e^{-0,017616j} \qquad (R^2 = 0.1149). \tag{19}$$

The variance around the estimated curve is substantial as is indicated by

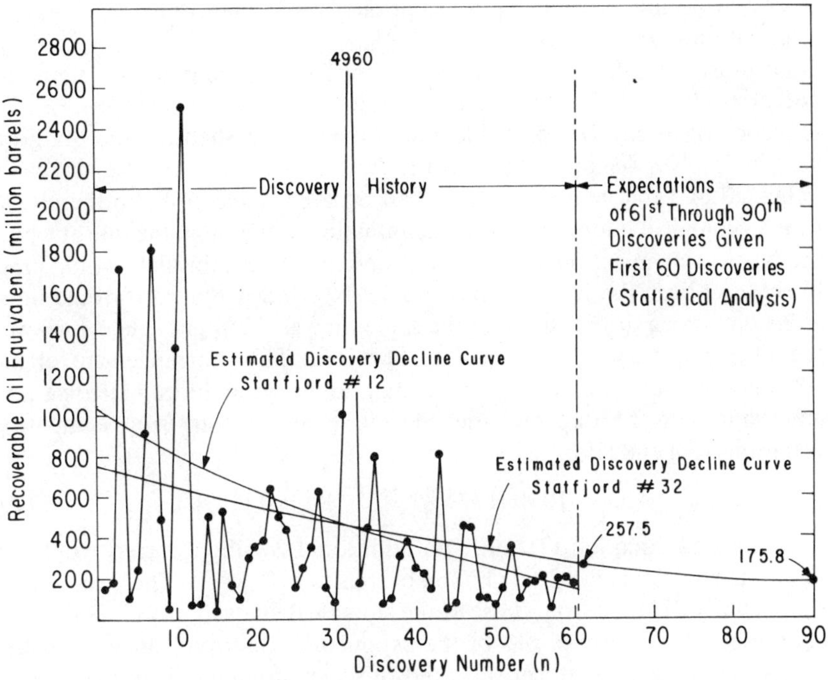

Figure 2. North sea-discovery history and forecast.

the low R^2 and is also apparent from Figure 2. Figure 2 shows the actual discovery history of the North Sea (Table 1), the estimated discovery decline relationships of equations (19) and (20) and the predicted sequence of expected discovery sizes of the statistical analysis. Figure 2 indicates that a naïve extrapolation of equation (19) will produce an expected discovery sequence similar to that of the statistical analysis.

According to the one-play analysis of the North Sea, using the statistical version of the disaggregated process model, there are 8.5 billion barrels of discoverable oil reserves in the North Sea at an average expected price of \$12 (MERS = 90 MMB) and 5 billion barrels at a \$9 price (MERS = 120 MMB) in addition to the recoverable reserves of the historic discoveries (7). Equation (19) indicates that in a \$12 world an additional 7.67 billion barrels of oil will be found. The price sensitivity of the exponential discovery decline function is lower than that of the statistical analysis due to the different location and slope of the exponential decline function. The steeper slope of the exponential decline function implies that ΔUR in Figure 1 will be smaller following an increase in price and a corresponding decrease in MERS and increase in N. The N deduced from the intersection of MERS and the discovery decline relationship of equation (7) is 17 percent higher than that of equation (19).

To illustrate some of the uncertainty inherent in the data as well as the potential effect of government regulation of the exploratory process, the sequence of discoveries of Table 1 was rearranged slightly. The Statfjord field in the Norwegian sector was the 32nd discovery in the North Sea as indicated in Table 1. If exploration had proceeded according to the rate at which geological information was accumulated, rather than the rate at which the Norwegian government chose to license its blocks, this field would most likely have been the 12th discovery in the North Sea. Statfjord would most likely have been drilled at about the same time as the Brent field, which was the 11th discovery and adjacent to Statfjord. This rearrangement of the sequence of discoveries resulted in a dramatic change in the estimated parameter values of the discovery decline curve as is clear from comparing equations (20) and (19):

$$D_j = 1031.91\, e^{-0{,}030161j} \quad (R^2 = 0.1129). \tag{20}$$

According to equation (21) only an additional 2 billion barrels would be found in a $12 world and 1.25 billion barrels in a $9 world. These results are significantly different from those of the statistical analysis as well as those of equation (19). The simplicity of the exponential discovery analysis makes it possible to easily perform this kind of sensitivity analysis of the process output with respect to the data base. Given the uncertain nature of the data base this is a desirable feature of a model.

3. A VALIDATION EXERCISE

The discovery decline relationship of equation (13) is a naive formulation of the geological depletion phenomenon. Because there is no satisfactory description of the process by which plays are being sought and discovered, the exponential decline relationship may serve as a first approximation to the geological depletion of a basin.

To indicate the forecasting ability of the exponential decline formulation equation (13) was estimated on the basis of the first 30, 40, and 50 discoveries of Table 1 and then used to "forecast" the next 30, 20, and 10 historical discoveries, respectively. Because of the significance of the location of Statfjord in the discovery sequence, the estimation/forecasting exercise was performed with Statfjord as the 32nd discovery and the 12th discovery, which an analysis of the process of generation of geological information supports.

When estimating equation (13) from the 30 first discoveries of Table 1 (Statfjord as the 32nd discovery is not included in the data base), the following relationship emerges:

$$D_j = 806.372\,e^{-0{,}0324}j.^5 \tag{21}$$

It is obvious from a comparison of equations (21) and (19) that an extrapolation of equation (21) will seriously underestimate the 31st to 60th discovery. The mean value of the actual discovery sequence (discovery no. 31 through 60 of Table 1) is 424.3 MMB whereas the mean of the predicted values is only 192 MMB.

If equation (13) is estimated from the first 40 discoveries the inclusion of Statfjord rotates the discovery decline relationship.[6] The forecasting bias introduced is changed from one of underestimation to one of overestimation. The mean value of the 41st to the 60th discovery is 212.5 MMB, whereas the mean of the predicted values is 535.6 MMB.

The overestimation bias is retained also when the number of observations in the data base is increased to 50.[7] The mean of the next 10 historic discoveries is 177.5 MMB. The mean of the next ten predicted values is 343.6 MMB.

The fact that regulation of the exploratory effort caused Statfjord to be found so late in the discovery process introduces estimation problems in any naïve analysis of the observed sequence of discoveries. The estimated discovery decline relationships improve considerably when Statfjord is move back to the 12th position in the discovery sequence. The 30 first discoveries then produce the following discovery decline relationship:

$$D_j = 1011.61\,e^{-0{,}028166j}. \tag{24}$$

The forecasting performance of the exponential decline relationship, equation (13), is increased dramatically when Statfjord is moved back to the 12th position in the discovery sequence. Table 2 reports a comparison of the actual and predicted discovery sequence using the first 30 discoveries to predict the next 30 discoveries.

The forecasting performance is good as measured by the low root mean square error (RMS), the standard deviation (std. dev.), as well as by the difference between the mean of the actual discovery sequence and the mean of the predicted discovery sequence. Although the individual differences are considerable, the mean of the predicted values is only 10.6 percent higher than the mean of the actual values in Table 2. In terms of additions to recoverable reserves, the upward bias of the predicted values is higher because the sub marginal discoveries of the actual discovery sequence would have to be removed before additions to the recoverable reserves base resulting from the 30 discoveries could be calculated. We would expect this bias to be alleviated as the more recent discoveries mature to the same stage in the extension/revisions process as the early discoveries.

In spite of the fact that equation (24) is statistically insignificant, the

Table 2. Forecasting North Sea discoveries 30/30 Statfjord # 12.

Order	*Actual discovery*	*Predicted value*	*Predicted minus Actual*
31	75.	422.	347.
32	1000.	411.	– 589.
33	178.	399.	221.
34	450.	388.	– 62.
35	800.	377.	– 423.
36	71.	367.	298.
37	100.	357.	257.
38	300.	347.	47.
39	375.	337.	– 38.
40	250.	328.	78.
41	220.	319.	99.
42	150.	310.	160.
43	800.	301.	– 499.
44	50.	293.	243.
45	75.	285.	210.
46	450.	277.	– 173.
47	450.	269.	– 181.
48	100.	262.	162.
49	100.	254.	154.
50	75.	247.	172.
51	150.	241.	91.
52	350.	234.	– 116.
53	100.	227.	127.
54	175.	221.	46.
55	175.	215.	40.
56	200.	209.	9.
57	50.	203.	153.
58	200.	197.	– 3.
59	200.	192.	– 8.
60	175.	187.	12.
Mean	261.47	289.25	27.78
RMS			220.40
Std. Dev.	234.95		222.38

forecasting error of cumulative reserves discovered is less than 11 percent even when the length of the predicted discovery sequence is as long as the sequence of observed discoveries from which the relationship was estimated. In the case of 40 observations[8] and a length of the predicted sequence of 20 the cumulative error is 13.5 percent, and in the $50/10^9$ case the error is 11.9 percent. The simplified "discovery model" of equation (13) seems to have some forecasting ability when applied to the cumulative discovery sequence of a basin as opposed to the sequence of individual discoveries.

Since the focus of the Basin Model is on the annual level of additions to reserves, the inclusion of the exponential discovery decline relationship in a Basin Model of oil supply seems to be a defensible proposition.

4. THE ELASTICITY OF OIL SUPPLY

The purpose of an economist's oil supply forecasting exercise is to construct a framework for integrating the knowledge of the geologist, the petroleum engineer, the economist, and the financial analyst so that the relationship between a set of economic and policy variables and the supply of oil can be identified and exploited. The economic variables are the exploration and production technology that determines the input factors required to identify, develop and produce a reservoir, the factor markets that determine the total cost of these input factors, the tax and regulatory regimes that impose constraints on some stage of the supply process, and the markets that determine the price of oil. These economic variables are embedded in the reservoir analysis referred to above. When discovery analysis is based on the level of recoverable reserves rather than the oil in place, economic variables are also embedded in the discovery decline relationship itself.

The ultimate impact on the flow of oil resulting from an increase in price or a reduction in cost is the sum of the impact on the activity in all the stages of the supply process. To avoid the more complex issue of the impact of an expected increase in the rate of change of the price of oil we will limit the discussion to the impact of an upward shift in the price of oil. The discussion refers to the behavior of a "price-taker" rather than a "price-maker," i.e., the behavior of a supplier that does not consider the impact of his decisions on the price level.

The immediate effect of a rise in the level of price (disregarding its impact on price expectations) is to create an economic incentive to "speed up" the recovery of a given level of reserves in a reservoir. An extensive discussion of this aspect of the elasticity of supply can be found elsewhere (1). The experience in the United States indicates, however, that the marginal cost of speeding up the production of a reservoir is a steeply rising function so that not much supply reponse has been observed following the recent jump in the price of oil. Price regulation must, however, take most of the blame for this poor response. Incorporating the speed-up effect in the reservoir analysis was considered too costly in terms of the increased complexity relative to the significance of this element of the elasticity of supply.

More significant is the fact that an increase in the price of oil enables an operator to invest in processes that will increase the recovery factor or that fraction of the oil in place that may be economically extracted (16). A shift in the recovery factor will shift the discovery decline curve of Figure 1 in a northeastern direction. Enhanced-oil-recovery processes (EOR) may require a different level of oil in place to be profitably installed from that of the oil in place reflected in the MERS analysis, assuming conventional recovery techniques. A specific MERS-relationship may thus be generated for each recovery process. Such a set of MERS-relationships would shift the discovery decline curve of Figure 7 in a piecemeal fashion. Because offshore reservoirs are produced primarily by conventional primary and secondary recovery technique and because our focus has been on offshore reservoirs, no explicit analysis of the economics of EOR has been included in the present version of the reservoir analysis. Onshore, however, EOR may constitute a significant element of the elasticity of supply as indicated by the NPC-study referred to above.

The expected monetary value of the reservoirs to be found, as well as their total number (due to the MERS and N effects), will increase with an increase in the price of oil as will the incentive to drill. The rate of exploratory drilling in the North Sea is regulated by the government through the licensing of individual blocks and the working program provisions of the concession agreements (5, 7). In the United States the schedule of lease sales regulates the access to drillable prospects. The government constraints on the level of exploratory drilling may be the reason that so little work on the optimal rate of drilling has been done. The Basin Model lends itself to analysis of the economics of the rate of drilling.

The interaction of the discovery analysis and the reservoir analysis allows us to estimate the price elasticity of the ultimate level of recoverable reserves, UR. The elasticity of UR is a function of how flat the discovery decline curve is at the MERS level, as illustrated in Figure 1. The elasticity of UR and hence also of the supply of oil will vary depending on the price range that is considered, the time horizon of the supply analysis, and the starting point of the analysis. The elasticity of supply of North Sea oil in 1985, given 1976 as a point of departure, was estimated at about .67 in the \$9 to \$12 price range (\$ mid-1976) as opposed to only half that value in the \$12 to \$15 range (7).

The largest share of total production comes from large reservoirs. The largest fraction of UR is in large reservoirs as indicated by Figure 1. The major determinant of the level of supply in a regulated world is therefore the government determined access to drillable prospects. This fact is often overlooked in the regulatory debate which is focused mostly on the issue of price. The inability to predict regulatory policy implies that the major determinant of the future level of oil supply is included in a judgmental if not arbitrary fashion in most studies of oil supply. The Basin Model may be used, however, to trace the implications of marginal changes in the level of regulated drilling activity. By identifying the elasticity of supply in a regulated drilling environment, the model may be used to approximate the likely supply response from a change in price control policy with or without an accompanying change in the policy with respect to the rate of exploratory drilling.

5. THE BASIN MODEL IN PERSPECTIVE

The North Sea exercise discussed above indicates that even if the exponential decline curve explained very poorly the individual historic observations, the ability to predict the cumulative reserves resulting from aequence of discoveries was good. When Statfjord was moved to the 12th location in the discovery sequence, the discovery function fitted to the 30 first North Sea discoveries overestimated the cumulative reserves resulting from the next 30 discoveries by only 10.6 percent. The overestimation may be explained by the fact that the earlier discoveries have graduated from the extension/revision process, whereas the reserve estimates of the more recent discoveries have not yet reached this stage of maturity.

The impact of government regulation of access not only to the basin as a whole but also to individual prospects in the basin was clearly demonstrated by the Statfjord case. The significance of knowing the most likely sequence was apparent from the sensitivity of the estimation/forecasting results to the location of Statfjord in the discovery sequence. The fact that the largest structure in the North Sea is lying on acreage held by the Norwegian government (12), and thus has not yet been drilled, makes one suspect that a naive extrapolation of the estimated discovery decline relationship may severely underestimate the level of resources to be discovered.

The North Sea discoveries of Table 1 may be subdivided into four plays. The first 30 discoveries include discoveries in all four plays. The exponential decline relationship was thus estimated from and used to predict the simultaneous depletion of four plays. Intuitively we would expect a simple exponential discovery decline relationship to fare well when the predicted discoveries are from the same plays as the discoveries over which the relationship was estimated. If additional plays are discovered we would expect the exponential decline relationship to underestimate the cumulative reserves of the predicted discovery sequence. The North Sea data-base did not allow us to demonstrate this bias when the discovery history of an area is naïvely used to forecast the future of a multi-play basin. This bias can be dealt with by letting the largest expected discovery size, L, reflect the basin extreme rather than an average of the largest historic discovery.

By combining discovery analysis and reservoir analysis, the expected number of oil reservoirs to be discovered can be determined as well as the level of ultimate recoverable reserves. Figure 1 visualizes the relationship between the geological and the economic assumptions entering this crucial indicator of the future level of supply.

The rate of exploratory drilling, the geological success rate, the reservoir development profile, and the reservoir production profile are exogenous to the present version of the Basin Model. These exogenous variables determine at what rate the recoverable reserves will be produced over time. These variables may be specified as endogenous variables in the Basin Model. Each of these variables must be endogenized to analyze all the elements of the elasticity of supply of oil. The recovery factor, the development profile, and the production profile must be endogenized to calculate the marginal cost of further development and production of existing reservoirs. The location and shape of that marginal cost relationship determines the supply elasticity of existing reservoirs as well as serving as a guide to whether further exploration should be undertaken as opposed to further development. The marginal cost curve of future reservoir development thus also enters into an analysis of the level of exploratory drilling.

Today the rate of exploratory drilling is determined as much by government policy as by the economics of exploration. Governments have slowed down the level of activity as compared to the level likely to have been produced by the economic incentive. Since taxes constitute a significant part of the costs as seen by the operator, the government may also provide further economic incentive by reducing taxes and thus reducing the MERS. From Figure 1, however, it is obvious that in terms of additions to the level of recoverable reserves and hence also to the supply of oil, access to a basin characterized by expected discovery sizes significantly larger than MERS is more important than marginal changes in the MERS itself.

The fact that there is an economic rent associated with exploiting the non marginal reservoirs is in general also the problem that causes the regulatory delay. The heterogeneity of the population of reservoirs makes it difficult to create a tax system that will capture the rent without distorting the activity. Doubts about the competitiveness of the oil industry and a desire to "control" the activity are factors that further complicate the problem of designing a "rent-capturing" system. When a problem reaches a certain level of complexity, ideology rather than problem-oriented thinking will take over. Since we are not very good at predicting the outcome of ideological debates, the degree of access or the level of exploratory drilling remains a source of coniderable uncertainty when forecasting the future level of oil supply.

Conceptually the Basin Development Model is a disaggregated process model (1, 8, 14). Rather than relying on a statistical analysis of the exploration process, the Basin Model relies on a simple exponential decline function to generate a discovery sequence within a sedimentary basin. The representation of the discovery process in the Basin Model is similar to the analysis made by geologists and engineers to provide quick estimate of the possible sizes/number of oil/gas fields in new basins (9). The Basin Model is as such neither a substitute for statistical analysis of the exploratory process nor a substitute for geological/engineering evaluation of specific prospects when the data and the resources to carry out more comprehensive analysis are available.

FOOTNOTES

The author is indebted to James L. Smith and Arlie G. Sterling for their excellent research assistance in many parts of this work, and to M. A. Adelman, R. S. Pindyck, and H. D. Jacoby for helpful comments. The work was supported by the Norwegian Research Council for Science and the Humanities and by the U.S. National Science Foundation under Grant No. SIA75–00739 as part of a larger MIT project to develop analytical models of the world oil market. However, any opinions, findings, conclusions, or recommendations expressed herein are those of the author and do not necessarily reflect the views of National Science Foundation.

1. A prospect or a concept may be defined as a geological configuration conceived to have trapped hydrocarbons that form a prospect for drilling.

2. A play is defined as a group of similar geological configurations, generated by a series of common geological events, forming a statistical population, and conceived or proven to contain hydrocarbons.

3. A basin is a continuous containment of possible hydrocarbon source rock.

4. The mathematics and numerical analysis techniques necessary to do these computations are beyond the scope of this paper [(see the work of Barouch and Kaufman (2)].

5. None of the R^2's of the equations estimated as part of this validation effort exceeded .1. The relationships are thus statistically insignificant. It is obvious from Figure 2 that the variance around any mean discovery function is bound to be substantial, rendering the relationship statistically insignificant. However, because of the ability to predict the cumulative additions to reserves rather than the individual discoveries, the forecasting performance of the estimated mean discovery relationships is reported below.

6. $$D_j = 630.62e^{-0.00324j}. \quad (22)$$

7. $$D_j = 722.259e^{-0.0134j}. \quad (23)$$

8. $$D_j = 1017.61e^{-0.028781j}. \quad (25)$$

9. $$D_j = 1026.56e^{-0.029663j}. \quad (26)$$

REFERENCES

1. Adelman, M. A., and H. D. Jacoby, "Alternative Methods of Oil Supply Forecasting," in R. S. Pindyck (ed), *Advances in the Economics of Energy and Resources*, Vol. II: *The Production and Price of Energy Resources*, JAI Press, Greenwich, Conn., 1978.
2. Barouch, E. and G. M. Kaufman, "Oil and Gas Discovery Modeled as Sampling Proportional to Random Size," MIT Sloan School of Management, Working Paper WP 888–76, December 1976.
3. ———, and ———, "Probabilistic Modeling of Oil and Gas Discovery," in *Energy: Mathematics and Models*, SIAM Institute for Mathematics and Society, 1976.
4. Beall, A. O., "Dynamics of Petroleum Industry Investment in the North Sea," MIT Energy laboratory Working Paper No. 76–007, June 1976.

5. Eckbo, P. L., "Planning and Regulation in the North Sea," *Northern Offshore*, No. 9, September 1976.
6. ———, *The Future of World Oil.* Ballinger Publishing Co, Cambridge, Mass. 1976.
7. ———, "The Supply of North Sea Oil," MIT World Oil Project Working Paper No. 77–015 WP, July 1977.
8. ———, H. D. Jacoby, and J. L. Smith, "Oil Supply Forecasting: A Disaggregated Process Approach," MIT World Oil Project, Working Paper No. MIT-EL 77–007WP, February 1977.
9. Ivanhoe, L. F., "Evaluating Prospective Basins," *The Oil and Gas Journal*, December 6, 13, 20, 1976.
10. Kaufman, G. M., "Statistical Analysis of the Size Distribution of Oil and Gas Discovery," in *Studies in Petroleum Exploration*, No. 1, AIME, 1975.
11. ———, Y. Balcer, and D. Kruyt, "A Probabilistic Model of Oil and Gas Discovery," in Studies in Petroleum Exploration, No. 1, AIME, 1965.
12. Lovegrove, M., "The Development of Oil and Gas Reserves on the UK and Norwegian Sectors of the Continental Shelf Off North-West Europe," paper presented to an OECD Workshop, November 1976.
13. Mannsveldt-Beck, F., and K. Wiig, *The Economics of Offshore Oil and Gas Supply*, Lexington Press, Lexington, Mass., 1977.
14. MIT World Oil Project, Supply Analysis Group, "Supply Forecasting Using Disaggregated Pool Analysis," MIT Energy Laboratory Working Paper No. 76–0009WP, May 17, 1976.
15. ———, "Progress on Analysis of the World Oil Market," MIT Energy Laboratory Working Paper, No. 75–015, October 1975.
16. National Petroleum Council, *Enchanced Oil Recovery*, December 1976.

ESTIMATING A POLICY MODEL OF U.S. COAL SUPPLY

Martin B. Zimmerman, MASSACHUSETTS INSTITUTE OF TECHNOLOGY

INTRODUCTION

It is becoming evident that traditional methods of econometric analysis are inadequate in estimating the long-run supply of minerals. Econometrics is limited to extracting estimates from observations of the past. Economic theory predicts that the least cost deposits will be exploited first.[1] The estimation of parameters from past data then must lead to biased projections of the future. Mining conditions in the future will be worse than in the past, and it is the impact of the future conditions on cost that we need to know.

In this paper we present a methodology for dealing with this problem

Advances in the Economics of Energy and Resources, Vol. 2, pp. 59–92.
Copyright © 1979 by JAI Press, Inc.
All rights of reproduction in any form reserved.
ISBN: 0–89232–079–6

We estimate a "cumulative cost" function that yields the long-run marginal cost of mining as a function of cumulative output produced over time.[2] This is relatively easy to do for coal mining in the United States, where exploration for new deposits is not a major factor in supply. The first step of the procedure relates cost to the rate of output and to the geological conditions of the mine. The second step combines geological information on remaining deposits with the cost function to yield the cumulative cost function. A central theme of both these steps is the interaction of economics with geology. The best deposits are selected first, and in each step of the estimation allowance must be made for this fact.

The problem of depletion is part of the general problem of structural change. The change through depletion in the nature of the exploitable mineral stock is only one factor that calls for an analysis sensitive to the changing circumstances of the industry. The factors impinging most heavily on the coal industry today are the strict environmental regulations legislated by Congress. Strip-mining and air-pollution standards limit the coal that can be supplied and burned. Air-pollution regulations were first introduced in 1970 and are still changing. National strip-mining legislation has just become law. Econometric estimation of supply functions based on past data would not be adequate to estimate the effects of these changing regulations. The modeling technique described below deals with this changing regulatory environment. Specific results are presented for Eastern and Midwestern deep-mining and for Western and Midwestern strip-mining. These results, in particular, have a bearing on current policy issues.

The analysis begins with a discussion of coal-mining technology. This provides the technical knowledge necessary for the cost estimation. After estimating cost functions for both strip mining and deep mining, the functions are combined with available data on remaining coal deposits to yield cumulative cost functions. The paper concludes with a discussion of the likely effects of environmental regulation on the evolution of coal costs.

TECHNOLOGY

There are two primary techniques for producing coal. Underground methods are used for coal seams lying deep (usually more than 150 feet) underground. Openings to the seam are constructed and mining takes place beneath the surface. When the coal seams lie close to the surface, the earth and rock above the seam, or the overburden, is removed and the coal is excavated at the surface.

Deep-Mining Technology

There are three types of deep mine: drift, shaft, and slope. When the coal outcrops on a hill, a drift opening is constructed that provides access at the

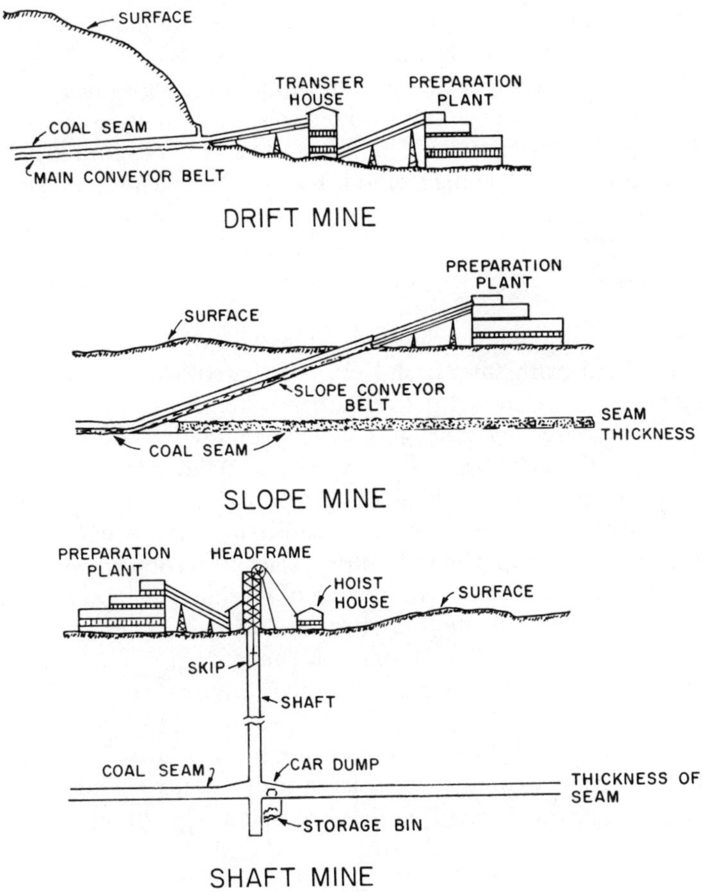

Figure 1. Principal types of coal mines. *Source:* S. M. Cassidy (ed.), *Elements of Practical Coal Mining*, Society of Mining Engineers, New York, 1973.

level of the seam. When the seam does not outcrop on a hillside, shafts or slopes are constructed (see Figure 1). Once the seam is reached, the mining of coal can begin.

The dominant technique for mining coal underground in the United States and the one upon which the following estimation is based is continuous mining. In continuous mining, a large machine rips the coal from the seam and loads it onto shuttle cars in one continuous operation. The shuttle cars transfer the coal to a central transport network for removal to the surface. A mining machine, two shuttle cars, and a complement of miners comprise a mining section. A mine consists of a number of sections, each working independently but sharing a common haulage system, ventilation system, and a set of openings to the seam that provide access for miners, for supplies, and a means for removing the coal.

Within a mining section, the ratio of capital to labor is relatively fixed. Noncontinuous techniques are also used, but at current factor prices the large new mine—the mine that determines the cost of coal at the margin—will use the continuous mining method. This will be true for a wide variation in factor prices.[3]

Surface-Mining Technology

In surface mining, the overburden is removed with large earth-moving equipment. Then, with smaller shovels, coal is removed and loaded onto trucks. After coal removal, the overburden is returned to the pit and the ground is recontoured and reclaimed. Equipment used varies from region to region in the United States. In the West and Midwest, where relatively large parcels of reserves lie under flat or gently rolling terrain, large draglines and shovels with capacities of up to 300 cubic yards are used. In the hills of Appalachia smaller equipment is the norm. The hilly terrain forces operations on narrow hillsides with frequent moving of machines. This can be accomplished only with smaller, more mobile machinery.

Surface mining is more capital-intensive than deep mining. The size of draglines and shovels has increased dramatically over time, so that one operator with one machine can move a large amount of material in one machine cycle. In the last five years, because of their great flexibility, large draglines have come to dominate the large-scale surface-mining operation.[4] The estimation below is based on dragline operation, although this is not important and consideration of shovels would lead to only minor changes.[5]

ESTIMATION PROCEDURE

Although the specifics differ, there is a common structure to the procedure used to estimate cost functions for both deep and strip mining. Mining costs are determined by the interaction of geology and of economic behavior. The rate of output is an economic choice and it is affected by the geology of the mine. This interaction is the subject of the following sections.

Physical Characteristics and Costs

The U.S. Bureau of Mines collects data on coal reserves classified broadly by seam thickness for deep reserves, and by the overburden ratio, that is feet of overburden to feet of coal seam, for strip reserves (13, 14). The latter ratio is simply a measure of the material that must be removed to expose a foot of coal seam. Clearly, these factors correlate with cost. However, one is impressed in reading a coal-mining manual by the multiplicity of factors affecting costs:

> Natural conditions involve roof, floor, grades, water, methane and the height of the seam.... In addition to these normal conditions, there are, in some mines, rolls in the roof or floor, and clay veins of generally short horizontal distance that intersect the coal seam. All these must be taken into account.
>
> It is possible for an experienced engineer to examine previous conditions of the sections and the immediate area of the section and assess proper penalties. As an example, if the roof is poor, production is reduced by as much as 15% of the available face working time. If the floor is soft, fine clay and water is present, the production handicap could be as much as 15%. If a greater deal of methane is being liberated, so that it is necessary to stop the equipment until the gas has been bled off, this delay could run as high as 10%. Fortunately, only a few mines in the United States have such severe conditions. The same remarks apply to all the other natural conditions. (1)

The following partial list of factors to be considered when planning a surface mine is from an authoritative mining manual.

1. Total coal reserves and general arrangement of deposit
2. Annual capacity
3. Nature of terrain
4. Maximum depth
5. Nature of overburden
6. Type of overburden, ... (8)

Unfortunately, data are not generally available on all these cost-determining variables. Ideally, the following cost function would be estimated:

$$c = \phi(G_1, G_2, G_3 \ldots G_n, Q)$$

where $G_i \equiv$ the N different appropriate geological characteristics and $Q \equiv$ output. For deep mining the only data available are on seam thicknesses. Assuming seam thickness is independent of other factors represented by ε, the following relationship is estimated:

$$c = \phi(Th, Q) + \varepsilon$$

where $Th =$ seam thickness
and $\varepsilon =$ disturbance term, reflecting other natural mining conditions.

Similarly, assuming overburden ratios are independent of other mining conditions, a strip-mining cost function is estimated as:

$$c = \psi(R, Q) + \eta$$

where R is the ratio of feet of overburden to feet of coal seam, and η a disturbance term. The disturbance terms reflect the nonobservable mining conditions. The more important and variable these factors are, the greater will be their variance. The variance, estimated together with the cost functions, will be directly incorporated in the long-run cumulative cost curves.

A Summary of the Procedure

No data are available detailing cost by mine. One is forced to rely on an indirect procedure. The above citations suggest a method for estimating cost. The production process within the mine is modeled as the sum of the operations of individual producing units. In deep mining, these are the sections. In strip mining the relevant unit is more complicated and is described below. The first step of the procedure is to estimate a relationship between productivity of the relevant producing units that comprise a mine and output and geological factors:

$$\frac{Q}{u} = f(Q, G_i) + \varepsilon \qquad (1)$$

where Q = output of the mine
u = the number of production units within the mine
G = observable geological characteristics
ε = unobservable geological characteristics.

Capital expenditures and expenditures for labor and supplies will be seen to be functions of the number of producing units:

$$E_i = f(u) \qquad i = 1, \ldots N \qquad (2)$$

where E_i refers to the class of expenditures. Finally, using equations (1) and (2), E can be expressed as a function of Q, G and ε. This then yields cost as a function of output, observable, and unobservable geological characteristics. While the procedure, in the abstract, is the same for deep and strip mining, the actual estimation differs significantly.

DEEP MINING COST FUNCTIONS

The Productivity Equation

The estimation begins with an analysis of drift mines where the depth is effectively zero.[6] Once the costs of a drift mine are determined, the added costs of deeper mining are considered.

The relevant unit in an underground mine is the producing section.

The productivity of a unit in isolation is written as:

$$q = A\,Th^{\gamma}\varepsilon \tag{3}$$

where q = output per section shift
Th = seam thickness
γ, A = constants
ε = disturbance reflecting the impact of unobserved mining conditions.

The multiplicative form for the equation expresses the interaction between natural mining conditions and thickness described in the above citation. One rarely observes a unit in isolation. The available data force one to measure q as total mine output divided by the number of unit shifts. A mine is usually made up of many units. Since these units share common equipment for haulage, ventilation, etc., their productivity is not independent of one another. Larger mines will have greater logistics problems. The larger the mine, the greater the travel time to the working sections. For these reasons, scale will affect the observed productivity per unit. To capture this effect, equation (3) is rewritten as:

$$\frac{Q}{S} = q = A\,Th^{\gamma}S^{\beta}\varepsilon \tag{4}$$

where S = the number of producing units, or sections
Q = total mine output.

Production per section should decline as congestion takes its toll. However, the problems faced by larger mines can be mitigated by capital expenditure. As congestion effects increase costs, new openings to the seam can be constructed, lessening the difficulties. In short, the observed scale effects will reach a limit as new shafts are opened to the seam. Equation (4) is rewritten to capture the effect of increasing the number of openings:

$$\frac{Q}{S} = A\,Th^{\gamma}S^{\beta}Op^{\alpha}\varepsilon \tag{5}$$

where $Op \equiv$ the number of openings (drift, shaft, slope) to the seam.

ECONOMETRIC DIFFICULTIES

Truncation

The fact that mining proceeds from the least costly deposits to ever more costly deposits presents a complication for estimation. The economic system generates the sample. Future mines cannot be observed. The observations consist of mines already opened. Since mines will open only if they can earn a profit, the sample consists of those mines with values of seam thickness and ε so that they earn a profit at current prices. For mines opening in thin seems the value of ε must be relatively favorable. This condition can be written as

$$A\, Th^{\gamma} S^{\beta} Op^{\alpha} \varepsilon \geq T$$

where T is the minimum productivity rate necessary for profitable operation. Therefore, while in nature Th and ε are assumed uncorrelated, in the sample this cannot be true, and therefore ordinary least squares (OLS) is biased. This is another manifestation of the inadequacy of past data in forecasting mineral supply.

The extent of the OLS bias depends upon how close the bulk of the observations are to the truncation point. In the United States east of the Mississippi River, the observations represent mines opened over many years. Most mines will be nonmarginal, that is, far from the truncation point. Consequently, the truncation of the sample is not likely to lead to a large bias. However, in order to eliminate this potential source of bias in OLS, a maximum-likelihood estimator is used that takes explicit account of the truncation problem.[7]

The Choice of T

The level of productivity sufficient for profitable operation is not known. An estimate of that level requires knowledge of the parameters to be estimated. The assumption used here is that the lowest productivity rate observed in the sample is the truncation point. This is an extreme assumption, since the actual rate could be lower. This assumption allows an estimate of the maximum bias of least squares.

Heteroscedasticity

The second problem arises because of the manner in which the data are collected. The data come from mine-inspection reports and production is the output of the entire mine. Recall that the productivity figure is derived by dividing total mine production by the number of sections in the mine. A problem in a large mine that shuts down one or two sections will have a relatively small effect on observed production per section. A small mine, on the other hand, could see production per section cut substantially. This means that the variance in production per section will be inversely proportional to the number of sections. Assume that the variance can be written as:

$$v(\log \varepsilon) = \frac{\sigma^2}{S}$$

where σ^2 is a constant and S the number of sections. The adjustment for heteroscedasticity thus simply weights all variables by the square root of S.

Results

The maximum likelihood results are as follows:

$$\begin{array}{lclll}
\sqrt{s}\,\log\frac{q}{s} & = & .7568\sqrt{s} + & 1.1071(\log Th)\sqrt{s} - & .2185(\log s)\sqrt{s} \\
\text{(s.e.)} & & (.4842) & (.1205) & (.0594) \\
\text{(t-statistic)} & & (1.5630) & (9.1906) & (-3.6762) \\
 & & +\,.0283(\log Op)\sqrt{s} & & \\
 & & (.0655) & & \\
 & & (.4314) & &
\end{array} \qquad (6)$$

$$\begin{aligned}
\text{S.E.R.} &= .9799\\
\chi^2 &= 232.4285\\
N &= 244.
\end{aligned}$$

These results indicate the importance of seam thickness and confirm our *a priori* expectations.[8] A 1 percent decrease in seam thickness leads to a 1.10 percent decline in productivity per section. The negative value of the coefficient of log s implies decreasing returns to the number of sections. The positive value of the coefficient of Op means, as expected, that additional openings to the seam stem the productivity decline. Also, the standard error of the regression, the estimate of σ, indicates a substantial dispersion about the conditional mean of productivity per section. The variance of productivity per section is:

$$v(\varepsilon) = e^{(2\mu + \frac{\sigma 2}{s}}\,(e^{(\sigma 2/S)} - 1)$$

where μ is the mean of the distribution of $\log \varepsilon$ and is zero by assumption. For a single unit:

$$v(\varepsilon) = e^{\sigma 2}\,(e^{\sigma 2} - 1) = 4.21.$$

This is a substantial dispersion and its implications are discussed below.

The productivity equation can be rewritten as:

$$q = 2.1314\,Th^{1.1071}\,S^{-.2185}\,OP^{.0283}. \qquad (8)$$

The number of units necessary to maintain a given rate of annual production in a given seam thickness, assuming three shifts per day and a work year of 245 days is:

$$S = \frac{\tilde{Q}}{\left(\frac{q}{S}\right) \times 3 \times 245} \qquad (9)$$

where $\tilde{Q}$ = annual output.

Solving equation (9) for S by substituting equation (8) in equation (9) yields:

$$S = \left[\frac{\tilde{Q}}{1566.579\ Th^{1.1071}\ OP^{.0283}} \right]^{1.2796}. \tag{10}$$

Underground Mine Expenditures

Once the number of sections is determined, the expenditures on labor, capital, and supplies can be determined. In addition to the capital equipment used in each producing section, a great deal of equipment is shared by all sections. This shared equipment includes ventilation equipment, transport for miners and machines, etc. The number of units determines the extent of the mine underground and should therefore determine the extent of this haulage, and ventilation equipment and support material such as rescue equipment.

This hypothesis is tested using engineering cost estimates prepared by the Bureau of Mines (11). These are estimates of the equipment, labor and supplies necessary for hypothetical mines. They represent estimates for mines assumed to produce a given level of output in a seam with a thickness either 48 or 72 inches thick under some unspecified set of mining conditions. Because the mining conditions are unspecified, it is difficult to assess how the number of necessary units was arrived at. These engineering estimates serve here only to establish the relationship between the number of units, however arrived at, and necessary expenditure.

The following equation was estimated for each class of expenditure:

$$E = \alpha + \beta S + \varepsilon$$

where E = expenditure
S = number of sections
ε = disturbance term.

The results are presented in Table 1. In spite of the small number of observations, the results are good. The high $\bar{R}^2$ indicates an implicit or explicit relationship used by the engineers in constructing these model mines.

The equipment expenditures are reported in equations (11) through (14). Capital equipment has been subdivided into equipment with different operating lifetimes. Thus, equation (11) represents capital goods with five-year lifetimes, equation (12) represents capital with ten-year lifetimes, etc. This breakdown is necessary since the correct treatment of depreciation must adjust for the useful life of the equipment.[9] Associated with the direct expenditures are expenditures for engineering, overhead, and various small construction tasks. These are assumed as a fixed 28 percent of initial direct capital expenditures by the Bureau of Mines (11). Operating and labor costs are shown as equations (15) and (16) of Table 1. Labor costs reflect

Table 1. Underground mining expenditures as a function of sections.

Equation	*Item*	*Constant*	*Slope*	R^2	*F*(1/5)	*S.E.R.*
(11)	I_5 5-yr capital	88380.5 (39528.5) (2.23587)	25528.0 (3102.92) (8.22708)	.9312	67.6843	46202.7
(12)	I_{10} 10-yr capital	2,193,350 (254037)	546842 (19941)	.9934	751.994	296930
(13)	I_{20} 20-yr capital	597,135 (36493.1) (16.3629)	53,870.7 (2864.65) (18.8053)	.9861	353.640	42,654.8
(14)	I_T Total capital	2,878,870 (210178) (13.6973)	626,241 (16498.6) (37.9571)	.9965	1440.75	245666
(15)	LAB Labor	407,608 (61761.2) (6.59974)	509,088 (4848.14) (105.007)	.9995	11026.5	72,189.2
(16)	SUP Supplies	−200,713 (94557.6) (−2.12265)	352,102 (7422.6) (47.4365)	.9978	2250.23	110523

Note: Numbers in parentheses are the Standard Error and t-Statistic, respectively.

union wage agreements. In addition, following the Bureau of Mines, 35 percent is added to account for overhead and fringe benefits.

Allowance is made for the union welfare charge (80 cents per ton in 1977), as well as indirect operating costs. The latter is estimated by the Bureau of Mines as 15 percent of total operating costs. The costs of loading the coal into trains must be added. The cost of a facility capable of loading up to 5 million tons per year is about $1 million. Annual labor costs for this facility add $57,000.[10] Finally, working capital, that is funds necessary to begin operations, is estimated by the Bureau of Mines as 25 percent of annual direct operating costs plus 33 percent of annual indirect operating costs plus 3 percent of initial direct investment.

In summary, the underground drift mine has been modeled as a conglomerate of producing sections. The productivity of these units was related to seem thickness, the number of sections, and the number of openings. In turn, expenditures were related to the number of sections.

Slope and Shaft Costs

For deep mines there are the added costs of sinking shafts and slopes, as well as the additional operating costs involved in mining deep underground. A recent estimate for a mine 796 feet below the surface indicates that a set of one intake and one outtake shaft cost $6,217,500 in 1974 (16). The cost of additional depth is small since fixed costs are large and the average cost per cubic foot of shaft declines rapidly as cubic feet increase. There is an additional ventilation cost and additional haulage costs, however the total is small.[12]

Costs in the Long Run

Summing equations, adding indirect costs, and the cost per shaft, or slope, and loading equipment yields total expenditures on equipment as a function of sections and openings. This is converted to an annual capital cost, based on the assumption of a 10 percent rate of return after tax, all equity financing, the sum-of-years depreciation and allowance for a 7 percent investment tax credit. The capital costs are added to operating costs, and all costs are adjusted for inflation to yield the total cost function in 1977 dollars:[13]

$$\text{TOTAL annual underground cost} = \$1{,}743{,}222 + \$2{,}122{,}480(\mathrm{S}) + 1{,}085{,}771\ (\mathrm{OP}).$$

Substituting equation (10) into the above equation yields:

$$\text{TOTAL annual underground cost} = \$1{,}743{,}222 + \$2{,}122{,}480\left[\frac{\tilde{\mathrm{Q}}}{1566.579\mathrm{Th}^{1.1071}\mathrm{OP}^{.0283}}\right]^{1.2796} + 1{,}085{,}771(\mathrm{OP}).$$

In the long run, the marginal cost of producing coal is the minimum average cost of a new mine. The task is thus to establish the minimum efficient scale of a mine, and evaluate the average cost at that point. Dividing by Q, differentiating with respect to Q and setting the resulting equation equal to zero, yields an expression for the minimum efficient scale:

$$\mathrm{Q}^* = \left[\frac{1{,}743{,}222 + 1{,}085{,}771(\mathrm{OP})}{593{,}445}\right]^{.7815} 1567\mathrm{Th}^{1.1071}\mathrm{OP}^{.0283}. \tag{18}$$

The minimum efficient scale depends upon OP, the number of openings, and upon Th, the thickness of the seam. The latter is determined by geology; the former is an economic choice. To determine the number of openings, we differentiate equation (17) with respect to openings, substitute equation (18) into the resulting equation, and solve for openings. The result, however, yields less than one opening as a result. Clearly, there must be a constraint. A minimum number of openings is necessary to produce. Two openings is taken as the minimum number, yielding an equation for minimum average cost:[14]

$$\mathrm{AC}^* = \frac{2567}{\mathrm{Th}^{1.1071}}. \tag{19}$$

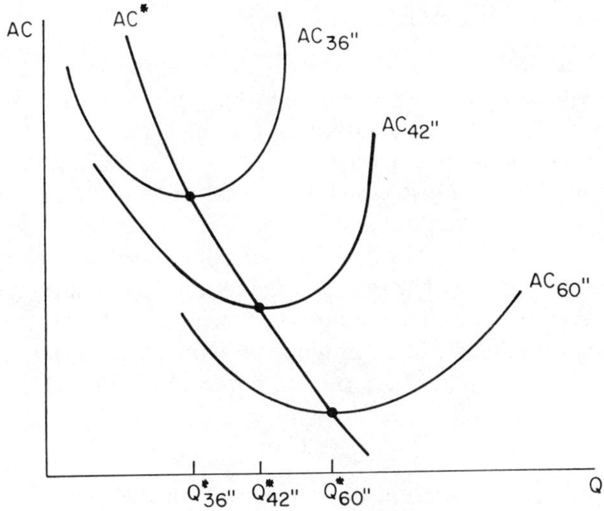

Figure 2. Locus of minimum average cost points—Ac*.

The resulting minimum average cost level of output is a function of seam thickness and ε. This is depicted in Figure 2 as locus AC*. The values of Q* shown in Table 2 indicate minimum efficient sizes for deep mines that fall short of the largest mines in actual operation. This is because the largest mines will represent some duplication of efficient units.[15] Also, the influence of ε means that for any given seam thickness, there will be a distribution of minimum efficient sizes. The locus AC* reflects the interaction of economics and geology. As seam thickness declines, the cost increase is mitigated, to some extent, by reducing the rate of output.[16]

Table 2. Minimum efficient size and Minimum average cost as a function of Th for deep shaft mines.

Th	*Q**	*AC**
28 inches	279,417	64.16
36 inches	369,051	48.58
42 inches	437,727	40.96
60 inches	649,674	27.60
72 inches	794,982	22.55
90 inches	1,017,762	17.62

Sources: Equations (18) and (19) where OP = 2, ε assumed equal to 1.

STRIP-MINING COSTS

The procedure followed in estimating the strip-mining cost functions is analogous to that used for estimating deep-mining cost functions. The analysis begins with the productivity of a mining unit, except that a unit is defined in a rather different way.

The dragline is the most commonly used equipment for overburden removal in large-scale strip mining. In the last five years it has dominated new equipment purchases. The cost functions developed below are therefore based on that technology. The size of draglines used in the industry has been increasing steadily over time. Since draglines have become larger and larger, the relevant unit is not the dragline itself but rather its size. Productivity is defined as the output, or overburden removed, per unit of size.

Size Unit

What is needed is a neasure of the capacity of the machine. Engineers put emphasis on what is called the maximum usefulness factor or MUF:[17]

$$\text{MUF} \equiv \text{d} \times \text{r} \tag{20}$$

where d = the capacity of the dragline bucket
r = the dumping reach.

These variables are shown in relation to a dragline in Figure 3. This is a measure, in effect, of the working capacity of the machine: How much material can be moved and how far can it be moved in one operation of the machine? How large a machine is necessary to remove a given quantity of overburden in a year? This will depend on how efficient a machine is. Efficiency can be expected to vary with size, since large size will affect how far a machine must move, and how long it takes for a machine cycle. These are

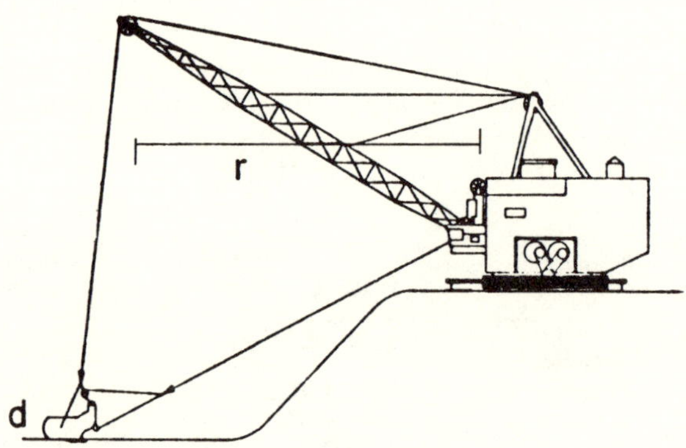

Figure 3. "Working diagram."

engineering relationships that are measured in the following regression:

$$MUF = A(OV)^{\alpha}\eta \qquad (21)$$

where OV ≡ total overburden removed, measured in cubic yards
η ≡ disturbance term reflecting natural conditions
A, α ≡ constants.

Again, η reflects the importance of characteristics that are not observed. It is reasonable to expect that this relation will also be multiplicative. Natural factors will interact with the quantity of overburden, making each cubic yard more difficult to move.

In many mines several draglines are used. The working capacity of two draglines each with one-half the capacity of a large dragline is equal, according to the above definition, to the capacity of one large dragline. However, in actual operation, using more than one dragline could be a more complicated process involving slower operations. As the overburden is removed from one part of the seam, the dragline moves to a new part of the seam. Draglines with larger reach need move less often than smaller draglines since their long reach allows them to uncover overburden over a wider radious. In order to capture this effect, equation (21) is rewritten as:

$$\Sigma MUF = A(OV)^{\alpha}N^{\beta}\eta \qquad (22)$$

where ΣMUF = the total capacity of all machines[18]
and N ≡ the number of machines in use.

The amount of overburden can be rewritten as a function of the product of the annual output, and the overburden ratio, R.

$$OV = ZRQ$$

where Z ≡ constant = .89 (see Footnote 19)
R ≡ overburden ratio ≡ feet of overburden per foot of coal
Q ≡ annual output rate.

Equation (22) is now written as:

$$\Sigma MUF = AZ^{\alpha}(RQ)^{\alpha}N^{\beta}\eta, \qquad (23)$$

the form actually estimated.[20]

Results

The resulting estimates are:

log MUF =	− .446684	+ .612306(log RQ)	+ .506967(log N)	
(Std. Error)	(2.31895)	(.145663)	(.229134)	(24)
(t-statistic)	(− .192623)	(4.20357)	(2.21254)	

$R^2 = .8595$
$F(2/7) = 21.4089$
S.E.R. $= .287762.$

The results confirm what one expects based on the history of increasing dragline size. The value of the coefficient of log(RQ) indicates there are increasing returns. It takes only a 0.6 percent increase in size to handle a 1 percent increase in overburden removal. The value of the coefficient of log N indicates that one large machine is better than two small ones having the same total capacity. This creates the incentive for larger and larger sizes.

There is also a fairly large dispersion, indicating the importance of natural factors. Because of the small number of observations, these results have been checked against more extensive data available from an earlier period.[21] In 1959, a survey of operating experience on draglines was conducted. The resulting equation for that period is:

log MUF =	− 5.01166 +	.858326(log OV) +	.711493(log N)	
(Std. Error)	(2.29250)	(.150932)	(.266627)	(24A)
(t-statistic)	(− 2.18611)	(5.68683)	(2.66850)	

$R^2 = .6161$
$F(2/21) = 16.8488$
S.E.R. $= .288071.$

While the parameters have changed, the character of the results is the same. The surprising result is that the estimated variance is also the same. The larger sample causes more confidence in the results, while the parameter values used are those for the 1975 data. Rewriting equation (24),

$$\text{MUF} = .6397(\text{RQ})^{.612306}\,\text{N}^{.506967}\,\eta. \tag{24B}$$

Expenditures

Following the procedure used for deep mining, expenditures are related to the number of producing units, in this case the size of the dragline. The size of dragline is a function of the amount of overburden to be removed. However, there will also be expenditures solely dependent on the amount of coal produced. Therefore, each class of expenditure is related to both the size of dragline and to the annual production of coal. An equation of the following form is estimated:

$$\text{E} = \alpha + \beta(\text{MUF}) + \gamma\text{Q}$$

where Q is annual output in tons. The results are presented in Table 3. Equation (25) is the dragline cost equation. This relates the cost to size. The data come from the major dragline manufacturers. Again, for calculating depreciation, capital goods are divided according to lifetime. The other

Table 5. Strip mine expenditures.

Equation	Item	Constant	Coeff. of MUF	Coeff. of Q	R^2	F	S.E.R.
(25)	Dragline cost	1,586,500	593.48		.9175	177.910	1,885,590
	(s.e.)	(702532)	(44.4945)				
	(t-statistic)	(2.25826)	(13.3383)				
(26)	I_T	3,107,220	17.2718	.213019	.9335	35.0953	222357
	(All capital						
	other than	(142884)	(5.72053)	(.0577415)			
	dragline)	(21.7465)	(3.01926)	(3.68919)			
(27)	I_5	268,374	−12.0284	.247652	.8680	16.4385	171719
		(110344)	(4.41771)	(.0445918)			
		(2.43215)	(−2.72274)	(5.55374)			
(28)	I_{10}	596,687	1.72287	.0112230	.4681	2.20031	65819.6
		(110344)	(1.69333)	(.0170920)			
		(14.1078)	(1.01745)	(.656623)			
(29)	I_{201}	1,762,480	22.0264	−.114750	.8461	13.7415	167271
		(107486)	(4.30334)	(.0434368)			
		(16.3973)	(5.11844)	(−2.64176)			
(30)	I_{20}*	479,674	5.55102	.0688944	.7736	8.54082	145369
		(93412.2)	(3.73987)	(.0377493)			
		(5.13502)	(1.48428)	(1.82505)			
(31)	LAB	448,156	−11.7455	.0427509	.8727	17.1455	107454
		(68894.9)	(4.40273)	(.0315682)			
		(6.50492)	(2.66778)	(1.35424)			
(32)	SUP	−61,161.9	62.0065	.228441	.9753	98.8775	237134
		(152040)	(9.71610)	(.0696658)			
		(−.402276)	(6.38183)	(3.27910)			

Note: Numbers in parentheses are the Standard Error and t-Statistic, respectively.
*I_{20} represents capital ineligible for the investment tax credit.

capital cost data are from hypothetical mine analyses of the U.S. Bureau of Mines (12).[22] Equation (25) suggests that MUF's do determine the cost of the dragline. The equation for all other capital expenditures also provides a good fit. When the expenditures are subdivided according to equipment lifetimes, the fit is less good. In two cases (equations 27 and 29) there is a negative coefficient. The fact that the toal investment equation is a good approximation, but the individual investment equations are not, indicates a good deal of substitutability among types of equipment. Some allocation of the total investment by life of capital good is necessary for depreciation accounting and equations (27) through (30) play this role here. About half of total capital expenditures are represented by the dragline so that incorrectly depreciating a fraction of the remainder of capital equipment will not lead to serious error. Comparing the strip expenditure equations to the deep expenditure equations indicates more flexibility in choice of equipment planning a strip mine. The other capital cost data are from hypothetical mine analyses of the U.S. Bureau of Mines (12).[22]

Costs in the Long Run

A total cost equation is constructed as for deep mining, by converting the capital expenditures to an amortized annual cost, adding labor and operating costs, and adding loading costs. Indirect costs are included on the same basis as they were for deep mining. The resulting cost equation is:

$$TC = 3{,}170{,}223 + 467{,}262\,MUF + .96Q. \qquad (33)$$

Substituting (24B) into (33), dividing by Q, and setting N = 1 yields:

$$AC = \frac{3{,}170{,}223}{Q} + \frac{299}{Q^{.3877}} R^{.6123}\eta + .96. \qquad (34)$$

Average costs are declining over the range in our sample. This suggests an unstable industry. However, while these economies of scale have theoretically not been exhausted, in reality there are always barriers in the short-run to building larger and larger equipment. Transportation of such large components presents a difficulty. Welding huge components might not be feasible. Therefore at any point in time average costs will turn up and the constraint is reached. Over time these constraints are overcome and the equipment increases in size. One constraint on the size of machine therefore is technological. However, there are other constraints operating on the observed size of machines. Large draglines are costly to transport over large distances. They are costly to erect and then to knock down. It is prohibitively expensive to erect the largest feasible dragline unless it will be amortized over a sufficiently large amount of cumulative production over time. In other words, only if the parcel of reserves is large enough will the technological constraint be binding. Otherwise, something smaller will be erected.

An important factor affecting the size of dragline actually chosen is the contour of the terrain. In very hilly terrain, only smaller machines will be stable. Also, in extremely hilly terrain equipment must be more mobile as mining proceeds from hill to hill. For all these reasons, the technological constraint is not likely to be binding in the majority of mines in a given area. If the technological constraint is substituted in equation (33), the resulting cost equation is a lower limit. Since nontechnological constraints are likely to be binding, these cost functions will be biased. Although data on machine sizes in new mines are unavailable, the available data can be used to estimate a nontechnological constraint.[24] Equation (24B) gives the relations between R, Q, η and MUF. New mines opening in a given area will push actual machine size to the constraint, $\overline{\text{MUF}}$. If one could observe η, along with R and Q, one would know $\overline{\text{MUF}}$. All that is actually observed is R and Q. Since η is lognormal, the best guess about $\overline{\text{MUF}}$ will come from substituting the geometric mean of the observed RQ in equation (24B), along with a value of one for η. The average cost equation results from inserting $\overline{\text{MUF}}$ into (33) and dividing by Q. Solving (24B) for Q after inserting $\overline{\text{MUF}}$ yields an equation for minimum efficient scale.

WEST (Powder River Basin):

$$\text{AC}^* = .52\text{R}\eta^{1.63317} + .96$$

$$\text{Q}^* = \frac{25{,}966{,}542}{\text{R}\eta^{1.63317}} \tag{35}$$

MIDWEST:

$$\text{AC}^* = .67\text{R}\eta^{1.63317} + .96$$

$$\text{Q}^* = \frac{16{,}277{,}052}{\text{R}\eta^{1.63317}}$$

Table 4 summarizes the above equations. The huge mine sizes predicted for very low overburden ratios are consistent with new strip mines opening in the western part of the United States. The greater open areas allow, in general, larger machines and lead to lower costs. The higher overburden ratios of the Midwest yield the smaller mines.

RESERVES AND CUMULATIVE COSTS

Equations (20) and (35) can be rewritten in log form as:

$$\log \text{AC}_\text{D}^* = \log \kappa_\text{D} - \gamma_\text{D} \log \text{Th} - \log \varepsilon$$

$$\log (\text{AC}_\text{S}^* - \gamma_\text{S}) = \log \kappa_\text{S} + \log \text{R} + \frac{1}{\alpha} \log \eta \tag{36}$$

Table 4. Minimum efficient size and minimum average cost as a function of overburden ratios for surface mines.

Powder River Basin

R	*Q**	*AC**
5	5,193,234	3.57
10	2,596,618	6.17
15	1,731,079	8.78
20	1,298,309	11.38

Source: Equation (35)
ε assumed equal to 1.
Costs shown exclude UMW welfare ton charge, cleaning costs, and reclamation costs under new law.

where D refers to deep and S to strip-mined coal. The parameter k_s depends upon the region, and is .52 in the West and .67 in the Midwest.

The distribution of underground coal according to the cost of mining depends upon γ_D, κ_D, the distribution of seam thickness, Th, and upon the distribution of natural factors, ε. The parameters γ_D and κ_D were estimated in equation (6). The variable ε is, by assumption, lognormally distributed. Its variance was estimated in equation (6) and the mean of log ε is zero. All that remains to be estimated is the distribution of Th.

Similarly, for strip mining, the parameters κ_S, γ_S and α were estimated in equation (24). The distribution of η is lognormal, with the mean of log η equal to zero, and variance as estimated in equation (24).[25] All that remains to be estimated is the distribution of R, or the distribution of strip coal according to the overburden ratio.

The Distribution of Th

The United States Geological Survey presents coal reserve data in broad seam thickness intervals. They record coal lying in seams between 28 inches and 42 inches thick and seams greater than 42 inches in thickness (13). Considering the impact of seam thickness on cost, this is inadequate for a complete description of the cumulative cost function. An approximation is used that establishes a method for more accurate estimation as more data become available.

There are data available on the distribution of tons of coal in the ground according to the thickness of the seam in which it lies for Pike County, Kentucky (26). This is a large and important coal-producing county in eastern Kentucky.[27] The object is to approximate this actual distribution by a well-known statistical distribution. The lognormal distribution has been used toward similar purpose in studies of other minerals and proves useful here (3).

Aitchison and Brown in their book on the lognormal distribution offer a test for lognormality (2). Plot the distribution on log-probability paper. If lognormality is a good approximation, the points will lie approximately along a straight line. This is done in Figure 4. Thus, for example, 50 percent of coal reserves lie in seams thicker than 42 inches. The fit is poorest in the tails, a common problem in work using the lognormal. However, the approximation appears adequate.

The variance is measured from the data shown in Figure 4. It is assumed that the lognormal distribution applies state by state and the variance remains constant from state to state. The mean in each state is estimated separately, using Bureau of Mines data on seams lying in seams thicker and thinner than 42 inches (13). The log of seam thickness distribution can be converted to a standard normal distribution:

$$\frac{\log \mathrm{Th}_i - \overline{\log \mathrm{Th}}}{\sigma_{\log \mathrm{Th}}} = u_{\log \mathrm{Th}} \tag{37}$$

where

$\log \mathrm{Th}$ = log of thickness i
$\overline{\log \mathrm{Th}}$ = mean of the logs of seam thickness
$\sigma_{\log \mathrm{Th}}$ = variance of the log of seam thickness
$u_{\log \mathrm{Th}}$ = the point on the standard normal distribution corresponding to $\log \mathrm{Th}_i$

or:

$$\overline{\log \mathrm{Th}} = \log \mathrm{Th}_i - \sigma_{\log \mathrm{Th}} u_{\log \mathrm{Th}_i}.$$

Using the percentage of coal in seams greater than 42 inches to establish

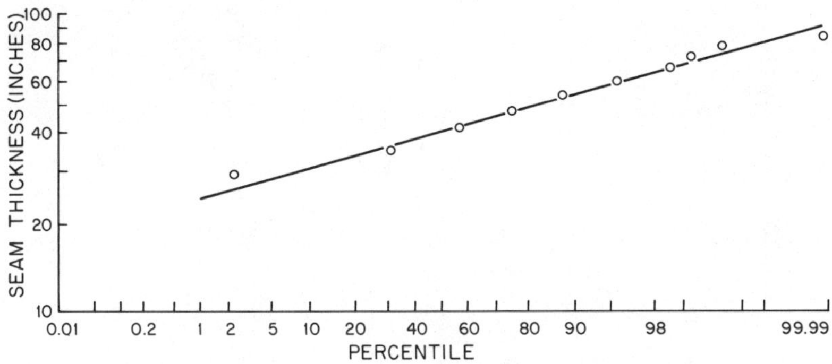

Figure 4. The distribution of coal reserves by seam thickness in Pike County, Kentucky.

$u_{\log Th_i}$, one can solve for the mean of the log of seam thickness in each state. Thus, if 50 percent of the coal lies in seams greater than 42 inches, $u_{\log Th_i}$ is the 50 percent point on the standard normal, or zero. Substituting the value of $u_{\log Th_i}$ allows one to solve for $\overline{\log Th}$. This procedure is repeated for each sulfur category, ranging from less than .4 percent to over 4 percent by weight. It is assumed that the sulfur is distributed independently of seam thickness.[28]

Since tons of coal by the log of seam thickness is distributed lognormally, the distribution of tons of coal in the ground according to the log of the cost of mining is the sum of two normal distributions and therefore itself lognormal.

The mean is equal to $\log \kappa_D - \gamma_D \overline{\log Th}$, and its variance is equal to $\gamma_D^2 \sigma_{\log Th}^2 + \sigma_{\log \varepsilon}^2$.

Strip Reserves

Again there is a paucity of data. There is information for the Powder River basin area of Montana, and for the State of Illinois on the distribution of coal according to the overburden ratio (7,9). These data are plotted on log-probability paper and shown in Figures 5 and 6. Again the fit is good over most of the range but diverges in the tail. Since the tails represent the highest-cost coal and in the Western region and Midwestern region this coal will not be used for hundreds of years, the lognormal approximation appears adequate.[29] The Montana distribution is assumed to hold

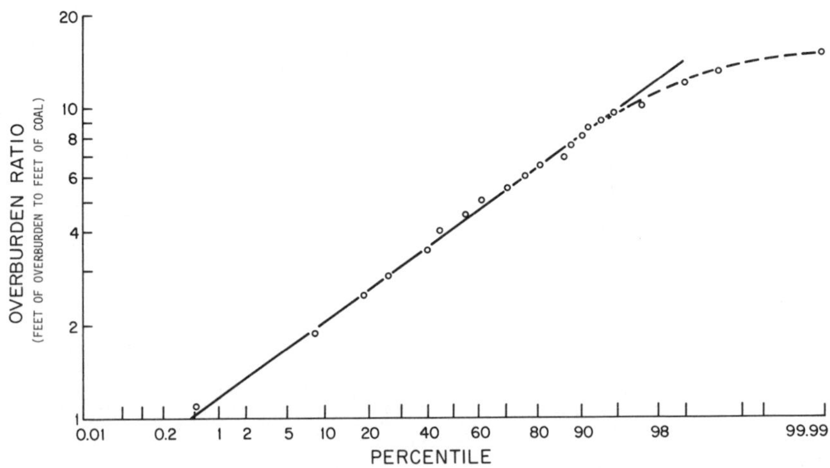

Figure 5. Distribution of strip mineable coal according to overburden ratio (Montana–Powder River Basin).

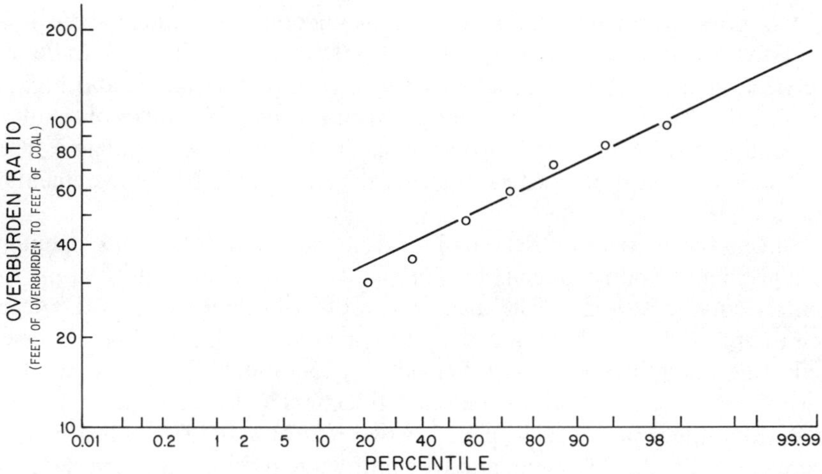

Figure 6. Distribution of strip-mineable coal according to overburden ratio (Illinois).

over the Western states. The Illinois distribution is assumed to hold over the Midwestern states.

The variances are estimated from the data in Figures 5 and 6. The means are calculated separately for each state. The Bureau of Mines has estimated how much strip coal lies at overburden ratios of less than what they call the economic limit and how much lies at greater ratios.[30] This economic limit varies from state to state. These data represent one point on the normal distribution function. These data are utilized to estimate the mean the same way that the mean of the seam thickness distribution was obtained. Since η and R are distributed lognormally, $(C\gamma_S)$ is also lognormally distributed with mean equal to $\log \kappa_S + \log R$ and variance equal to $\sigma^2_{\log R} + (1/\alpha)^2\sigma^2_{\log \eta}$.

TRUNCATION AGAIN

There is one remaining complication. The source of the difficulty is again the fact that the least-cost deposits are mined first. That implies that the coal remaining in the ground must be at least as costly to exploit as today's long-run marginal cost. If the distribution of coal according to cost is c, and if today's long-run marginal cost is $\bar{c}$, then the distribution of coal according to the log of the cost of mining can be written as:

$$\frac{\phi(\log c)}{1 - \int_{-\infty}^{\log \bar{c}} \phi(\log c)} \tag{38}$$

where $\phi(\log c)$ is normal. This is simply the truncated normal distribution.

The previous analysis indicates that in underground mining there is a trade-off between seam thickness and ε. For strip mining, there is a trade-off between R and η. At any constant level of cost, thinner seams and higher overburden ratios are compensated for by more favorable values of ε and η. If ε and η were observable, it would be a simple matter to calculate $\bar{c}$. The last mine that must be opened to satisfy demand would be the incremental mine.

Since ε and η are not observed, $\bar{c}$ must be estimated. New mines opening in a given area and producing a given quality of coal should all be approximately equal-cost mines. The distribution of seam thicknesses of the deep mines and of the overburden ratios of the new strip mines will be disperse, reflecting the influence of the other natural conditions. This is, in fact, the case. Data were collected on the seam thickness and overburden ratios of new deep mines in the east opened between 1973 and 1975 inclusive. The sample was divided by region and by sulfur content. In the Western states, a data base was available indicating seam thicknesses and overburden ratios for mines now in the planning stage.[31]

The data are summarized in Table 5. The wide dispersion confirms the earlier statistical estimates of the importance of ε and η.

The remaining problem is to estimate $\bar{c}$, given these observations. This involves selecting a value of Th to characterize the underground mine data

Table 5.

Area/sulfur content/type of mine	*Number of Observations*	*Range*	*Geometric Mean Th or R*	$\bar{c}$
	Deep Mines			
Northern Appalachia				
High sulfur (> 1.5%) (Shaft)	7	42–85	56.8	29.96
High sulfur (> 1.5%) (Drift)	10	36–78	53.3	27.54
Southern Appalachia (Drift)				
Low sulfur (<.8%)	29	35–72	44.1	33.25
Medium sulfur (.9 to 1.5%)	16	31–131	49.6	29.20
High sulfur (> 1.5%)	17	39–108	53.2	27.00
Midwest				
High sulfur (Shaft)	5	49–87	74.1	21.84
	Strip Mines			
Midwest				$c - \gamma_s$
High sulfur	4	7.5–22	17.0	12.46
Montana-Wyoming				3.76
Low sulfur	23	.07–11.5	5.2	

Source: Equations (19) and (35) in text.

N.B.: These costs exclude UMW welfare fee, coal-cleaning costs and state taxes. In strip mining they exclude reclamation costs. Drift mine costs are calculated excluding shaft costs.

and a value of R to characterize the surface mine. The underground cost estimate will then be given by $\kappa_D/\tilde{Th}$ where κ_D is the constant from equation (19) and $\tilde{Th}$, the chosen value of Th. Similarly, the strip mine cost will be given by $\kappa_s\tilde{R} + \gamma_s$ where κ_s and γ_s are constants estimated in equation (35) and $\tilde{R}$ is the chosen R. Again, given the lognormality of the distribution, the cost estimate that maximizes the likelihood of observing the distributions of Th and of R is obtained using the geometric means of the observed Th and of R respectively. These values are reproduced in Table 5.

Table 5 indicates that the incremental cost of coal varies inversely with sulfur content, and that high sulfur midwestern coal is less expensive than high sulfur eastern coal. These results are confirmed by market relationships. The lower strip-mining costs understimate the cost of strip-mined coal. However, an important item excluded from Table 6 is reclamation cost (see Footnote 22). The addition of reclamation cost would narrow considerably the difference between the estimate of $\bar{c}$ for midwestern strip-mined coal and for the deep coal. The cost of Western coal in Table 5 is also low. Actual price will include state taxes, which are high in Montana. Nevertheless, it appears there is a downward bias in the strip cost estimates. In the discussion that follows, we deal only with percentage increases in cost for strip mining and for deep mining separately. Consequently, the underestimate of the absolute cost level does not affect the results.

LONG-RUN COSTS

We are now able to calculate the distribution of coal in the ground according to the cost of exploitation. The distribution is given by equation (38). The amount of coal available at up to 10 percent more than current cost in a given state is the total state tonnage multiplied by the probability that the cost of production is between $\bar{c}$ and $1.10\bar{c}$:

$$T_i \int_{\log \bar{c}}^{\log 1.10\bar{c}} [\phi(\log c)/1 - \int_{-\infty}^{\log \bar{c}} \phi(\log c)dc]dc \tag{39}$$

where T_i is total state tonnage.

The data on the tons of coal in the ground come from the U.S. Bureau of Mines (13, 14). In addition to thickness and depth intervals, the Bureau divides coal reserves into certainty categories. The certainty categories are called measured, indicated and inferred. The first two are estimates based on observations at most $1\frac{1}{2}$ miles apart, due to mining or drilling. Beyond that point the reserves are inferred based on extrapolation. The Bureau of Mines calls all coal in the measured and indicated categories, lying in seams thicker than 28 inches and no more than 1,000 feet below the surface, the demonstrated reserve base. Only coal meeting these physical limits is considered technically mineable using present technology. The Bureau further divides the total into coal exploitable by underground methods, and coal exploitable by surface methods. The vast amount of coal in the demonstrated reserve base has been used to justify assumptions of perfect elasticity for long-run coal supply. In the East, the area of interest here, past mining and drilling has been extensive, meaning that most of the technically mineable coal will be in the measured and indicated categories. In the West, the numbers are probably an underestimate because previous activity has not been extensive. However, given the great elasticity of supply in Western coal seen below, this is not a serious bias. Consequently, we follow previous practice (15) and use the demonstrated reserve base.

Equation (39) defines implicitly the cumulative cost function. If a cumulative output total is specified, equation (39) can be solved for the upper limit of integration. The upper limit will be the marginal cost of mining resulting from having mined that volume of coal.

The results of this analysis are summarized in Tables 6 and 7. Table 6 presents estimates of how fast costs would rise in Eastern underground mining if current rates of underground output were maintained for five years, ten years, thirty years and fifty years. Thus, at current rates of output it is estimated it would take twenty years for the costs of high-sulfur coal to rise 10 percent in northern Appalachia. In southern Appalachia high-sulfur coal costs are estimated to rise by 11.4 percent in twenty years. A particularly interesting result is the greater elasticity of Midwestern deep output compared to Midwestern strip production. This implies that we can expect to see an increase in the relative share of deep output in that region, a reversal from previous history. However, planned new mine openings, in fact, confirm this result.[28]

These estimates put the popular interpretation of "reserves" into perspective. Popular impressions as well as authoritative estimates have used the large amount of coal in the reserve base to justify the assumption of no increase in costs over time due to depletion.[32] In fact, the estimated cost increase for high-sulfur coal are not dramatic, but it is far from negligible. The increase in low-sulfur costs is significant.

Tables 6 and 7 sum up the central policy dilemma with respect to the large-scale substitution of coal for liquid and gas fuels. Environmental regulations limit emissions of sulfur dioxide. This had led to a large demand for coal low in sulfur content. As Table 6 indicates, low-sulfur Eastern coal is in relatively inelastic supply. This had led to the great interest in the low-sulfur Western coal. Table 7 shows how greatly Western output can be expanded at a cost close to current levels. In the absence of sulfur regulations, Eastern high-sulfur coal could expand. In particular, Table 6 indicates Midwestern deep coal would likely expand and capture new markets. Sulfur regulations then increase the attractiveness of Western coal. Yet, the

Table 6. Estimated percent increases in cost over time at current rates of output.

	5 Years	*10 Years*	*20 Years*	*30 Years*	*50 Years*
Northern Appalachia					
medium sulfur	4.5	8.6	17.0	25.0	41.0
high sulfur	2.3	5.2	10.0	15.0	24.9
Southern Appalachia					
low sulfur	7.9	15.0	31.8	51.2	98.6
medium sulfur	6.9	13.2	25.7	40.9	68.0
high sulfur	2.9	5.9	11.4	16.8	29.6
Midwest					
high sulfur	0.6	1.5	2.7	4.1	6.9

Source: Text.

movement to Western coal involves another environmental problem. Western coal in large part, must be strip mined. In the arid climate of these Western states, reclamation of the land is a difficult task. It is uncertain what happens once the fragile ecology of the region is disturbed. This uncertainty has led some to question the advisability of developing Western coal. Others, fearful of the change in life-style accompanying large-scale mining operations, also wish to retard development.

The results of this paper indicate that there is a policy choice that must be made. The output of low-sulfur coal can be increased, but at the expense of rapid development of Western coal. This is a particularly difficult trade-off to be made since the costs of each action are borne in different regions. Air pollution in the cities can be reduced, but at the expense of strip-mining in rural areas. Only the advent of a low-cost desulfurizing device can slice through this gordian knot.

These environmental trade-offs are made more expensive by the desire to increase the use of coal as a substitute for oil and gas. An oft-repeated goal is to double the United States' production of coal. This would mean that

Table 7. Estimated percent increases in cost over time—Midwestern and Western surface-mined coal at current rates of output.

	5 Years	*10 Years*	*20 Years*
Midwest			
High sulfur	10.1	15.6	25.7
Montana-Wyoming*			
Low sulfur	1.3	2.5	8.2

Source: Text.

*These figures represent cost increases at five times current output levels.

costs would rise at twice the rate shown in Tables 6 and 7. Doubling output and retarding the development of Western coal would put severe pressure on Eastern low-sulfur coal. The country might not be willing to bear the price.

Other Factors Influencing Costs

This paper has dealt with only one element leading to increased costs: depletion. Depletion interacts strongly with environmental regulation so that the regulations we choose will, in part, determine the evolution of coal costs. Other factors will also importantly determine coal prices.

The Health and Safety Act of 1969, which specified safety regulations, has been cited as the cause of the decline in productivity experienced since 1970. The estimation of this paper relies on 1975 data, so that some of this impact has been incorporated. However, the decline has continued since 1975. As a first approximation this can be treated as a negative rate of technological change, causing the entire distribution of coal according to exploitation cost to shift to higher cost levels. How long this will continue, no one knows. In the long history of mineral supply technological change has worked to stem the cost increases imposed by nature. The technological regress of recent years can be overcome with new techniques.

Finally, the prices of input factors will affect costs over time. The coal industry itself is competitive, but important inputs are not competitively supplied. As oil prices increase, the United Mine Workers union will be in a position to capture some of the rents created.

SUMMARY

This paper has developed a method for dealing with structural change in a mineral industry. Depletion and environmental regulations are continually changing the supply responsiveness of the coal industry in the United States. Traditional econometric analysis is inadequate in the presence of this structural change. To deal with depletion, this paper has based its estimates of long-run supply upon the geology of remaining deposits. To deal with environmental regulations, cumulative cost curves were estimated separately by type of mining and by type of coal. The evolution of coal costs will depend crucially upon policy choices. There are, however, other factors this paper has not discussed, that will greatly affect the cost of coal. Finally, technological change over the truly long run will have the greatest impact on the evolution of coal prices.

FOOTNOTES

I am indebted to Michael Baumann and Chris Alt for research assistance *par excellence*, and to Jerry Hausman, who provided his program for truncated estimation.

Financial support of the U.S. Energy Research and Development Administration is gratefully acknowledged.

1. The theory of depletable resources predicts that lower cost resources will be exploited first. This has been discussed frequently in the literature. See, for example (4), for a discussion of behavior in a competitive industry. Similarly, it would never pay a monopolist to move a unit of lower-cost production to a future period in exchange for a high cost in an earlier period since the present value of profit stream would be reduced. Unless all is produced in a single period, any optimal set of prices and outputs will exploit the lower-cost deposits first.

2. The cumulative cost curve first appears in Hotelling's theoretical analyses of exhaustible resources (5). Hotelling considered the problem of the behavior of price over time when cumulative output led to cost increases. It can be shown that the following equation describes the behavior of price over time in a competitive industry:

$$rp = \left[\frac{dx}{dt} \times \frac{dc}{dx}\right] + \frac{dp}{dt}$$

where r = rate of interest
where p = *net* price (gross price minus cost)
x = cumulative output
c = marginal cost
t = time.

In this expression, $\frac{dc}{dx}$ is the cumulative cost function.

3. A unit is defined here as a mining machine, two shuttle cars, and a complement of miners.

There are no data on person-hours worked on production sections for individual mines that would allow a test of this constant capital-to-labor ratio assumption. However, engineering data do support the contention. Comparing descriptions of continuous mining operations in the 1950s with estimates of the Bureau of Mines in 1970 and with estimates in the *Mining Engineering Handbook* of 1973, indicates little variation in the composition of a continuous mining unit over a period during which relative factor prices were changing. Before the Safety Act of 1969, miners in a unit varied from seven to nine. See U.S. Bureau of Mines Information Circular, IC 7696, September, 1954. American Institute of Mining Engineers, *Mining Engineering Handbook*, Summer, 1973, and U.S. Bureau of Mines Morgentown Research Center, *Economic Evaluation of Underground Bituminous Coal Mines*, Report No. 70–10–A, September, 1973.

The continuous mining machine has been in use since the 1950s and the dominant technique in new mines since the early 1960s. In other words, although relative factor prices have varied in this period, continuous mining has still been preferred.

4. Draglines have increased steadily in size. In the 1940s the largest bucket was 25 cu. yards. By 1965 it had reached 100 cu. yards and today it is over 220 cu. yards. There was a concomitant rise in dumping radius (8).

5. Draglines have come to dominate large-scale shipping. One major manufacturer informed us that there have been no orders for large stripping shovels in the last five years. A complete analysis could treat other types of overburden removal equipment, but the results would be only slightly altered.

6. The data represent deep mines producing more than 100,000 tons per year. Smaller mines generally respond to spot market signals, entering and exiting as spot prices change. The smaller mines do not compete in the long-term contract market and represent a different production function. The data on seam thickness, number of openings and number of sections and shafts worked per day, days worked per year, and

output come from mine inspection reports of the Mining Enforcement Safety Agency.

In order to be sure that ignoring depth in the productivity equations was not a difficulty, an earlier data sample was used to test for the impact of depth. Data on a sample of mines for 1954 collected by the Bureau of Mines allowed a test. The results indicated that the null hypothesis that depth had no effect on productivity could not be rejected at a 50 percent level of confidence. There was no significant change in the coefficients or in the standard error of the regression when depth was added to the independent variables.

7. $\log \varepsilon_i$ is normally distributed by assumption. Therefore the probability of $\log \varepsilon_i$ is given by the following:

$$f(\log \varepsilon_i) = 0, \text{ if } \log \varepsilon_i < \log T - \log A - \gamma \log Th - \beta \log s - \alpha \log OP$$

$$= \frac{\phi(\log \varepsilon)}{1 - \int_{-\infty}^{\log A - \gamma \log Th - \beta \log s - \alpha \log OP} \phi(\log \varepsilon)}$$

$$\text{if } \log \varepsilon_i \geqq \log T - \log A - \gamma \log Th - \beta \log s - \alpha \log OP.$$

Jerry Hausman kindly provided the computer program for performing this estimation. For a discussion of this problem see Hausman, J., and D. Wise, "Social Experimentation, Truncated Distributions and Efficient Estimation," *Econometrica*, forthcoming.

8. The results also confirm expectations about the bias of OLS in this case. The maximum bias was, in fact, very small. The results are available from the author.

9. The depreciation rule used in these calculations is the sum of years digits. This is the most favorable method of depreciation for a company. See Hall, R. E., and D. W. Jorgensen, "Tax Policy and Investment Behavior," *American Economic Review*, June 1967, pp. 391–414.

10. TRW, Inc., *Coal Program Support Report* (prepared for the Federal Energy Administration), June 28, 1973, Figure 3–5A.

11. For confirmation see McGuire, E. J., "Do-it-Yourself of Simplified Shaft-Sinking Cost Estimates," *Coal Age*, February 1969, pp. 92–96.

12. To haul a ton of coal up 1,000 feet requires 1.3276 kwh. If the efficiency of the engine is 80 percent, the effective power needed is 1.66 kwh. At one cent per kwh this is a trivial amount. Ventilation costs increase according to the following formula:

$$HP = \frac{KOV^3(l)}{33{,}000}$$

where HP = increase in horsepower
O = area of shaft (24)
V = velocity of air (774 cu. ft. per minute)
l = incremental length of airway (2,000 ft)
K = coefficient of friction (2×10^{-8}).

This yields an additional power cost of $5,540 per year, again a trivial amount.

13. Cost of capital is calculated in the following way:

$$I = \int_0^T (c - \text{taxes}) e^{-rt} dt$$

where I = initial investment
c = annual return necessary to realize an after-tax return of r percent
T = life of capital good.

Let u = corporate income tax
v = depletion allowance in percent of gross profit, total depletion = v(c − depreciation).

Then:

$$\text{taxes} = u(c - \text{depreciation} - \text{depletion})$$

$$I = \int_0^T [c - u(c - \text{depreciation} - \text{depletion})]e^{-rt}dt$$

$$= \int_0^T ce^{-rt}dt - \int_0^T uce^{-rt}dt + \int_0^T u(\text{depreciation})e^{-rt}dt$$

$$+ \int u(\text{depletion})e^{-rt}dt$$

$$= cF - ucF + u\int_0^T(\text{depreciation})e^{-rt}dt + u\int_0^T v(c - \text{depreciation})e^{-rt}dt$$

where $F = \int_0^T e^{-rt}dt$

$$I = cF - ucF + uvcF + (u - uv)\,(\text{presented value of depreciation}) \qquad (1)$$

The last term on the right hand side is

$$(u - uv)I\left\{\frac{2}{rt}\left[1 - \frac{1}{rt}\left(1 - e^{-rt}\right)\right]\right\}$$

(see Hall & Jorgensen, *Op. cit.*, 1967.)

Substituting the above expression into equation (1) yields:

$$\frac{c}{I} = \text{cost of capital} = \frac{[1 - (u - uv)(1 - F/t)2/rt]}{F(1 - u + uv)}.$$

Costs were adjusted to 1977 prices using the BLS construction machinery wholesale price index. Wages were adjusted by the income in the BLS bituminous coal-mining average wage.

14. The choice of two openings reflects the minimum number observed in the sample, and what is necessary for intake and outtake.

15. See, for example, the description of the Wabash Mine, *Coal Age*, September 1974, p. 102, for an indication of duplication of units.

16. For a discussion of the implications of this analysis, see Zimmerman, M. B., "Modeling Depletion in a Mineral Industry—The Case of Coal," *Bell Journal of Economics*, Spring 1977, pp. 54–55.

17. The use of MUF numbers was introduced by H. Rumfelt in "Computer Method for Estimating Proper Machinery Mass for Stripping Overburden," *Mining Engineering*, Vol. 13, No. 5, May 1961, pp. 480–487. This is summarized in (9).

18. Some mines were using shovels and draglines. The comparable MUF number for a shovel were added to those of the draglines. The adjustment for shovels simply considers the equivalent dumping reach. This is discussed in (17), p. 420.

19. The value of Z is derived in the following way:

$$\text{TOTAL Cubic Yards of Overburden} = \frac{\text{Feet of overburden} \times \text{acres mined} \times 43{,}560}{27} \qquad (1)$$

where 43,560 = square feet in an acre
27 = cu. ft in a cu. yd.

But: (2)

$$\text{Acres mined} = \frac{Q}{Th \times 1800} \tag{2}$$

Where Q = annual output
Th = thickness of seam in feet
1800 = tons per acre – foot

Substituting equation (2) into equation (1) yields:

$$\text{Total cu. yds} = \frac{\text{ft of overburden} \times (Q/Th \times 1800) \times 43}{27}$$

$$= .89\,QR.$$

20. The data on production and overburden ratios are from State of Illinois, *Department of Mines and Minerals, Annual Report 1975.* The information on equipment size is from direct correspondence with the companies involved. Truncation made little difference in the estimates. The OLS estimates are reported here.

21. The earlier data are from (17), p. 408.

22. The sample excludes the contour mines, since they represent a different technology. Also, recent estimates for three mines are available, but they include equipment and labor for reclamation cost. The idea here is to exclude reclamation cost so that the impact of new reclamation legislation can be added on separately. The data used here represent practice as of 1969, when reclamation requirements were not nearly as stringent as at present.

23. For an example of this same phenomenon in the petrochemical industry, see Adelman, M. A. and M. Zimmerman, "Prices and Profits in the Petrochemical Industry," *Journal of Industrial Economics*, June 1974, pp. 245–254.

24. The best approach would be to establish a relation between contour, parcel size, etc., and the size of dragline chosen. Then, reserves would be classified along these additional dimensions. Unfortunately, the requisite data do not exist.

25. If the mean is not zero, the constant term will incorporate the bias.

26. The data include coal that is recoverable with present underground techniques, that is, coal lying in seams at least 28 inches thick. Thus, the distribution should be interpreted as the distribution of coal that is mineable with today's techniques. Exploiting seams thinner than 28 inches would involve using other techniques and would lead to a large discontinuous jump in the cost function. Excluding the thin coal as we do here means that the upper tail of the cost distribution is somewhat understimated. More coal is available at very high costs. However, this does not change the qualitative results. Furthermore, exclusion of this very thin coal is consistent with the Bureau of Mines' Reserve Base measure, discussed below, as well as with the FEA's 1974–1976 work on coal supply. The data are from: Paul Weir Company, *Economic Study of Coal Reserves in Pike County, Kentucky and Belleville District, Illinois* ..., Job No. 1555, January 1972. The data from Belleville District consider reserves in only one seam and thus are too restrictive to establish a distribution.

According to the United States Geological Survey *Bulletin 1120*, 1963, past mining had depleted about 12 percent of the original reserves lying in seams greater than 28 inches thick in Pike County. This means that the present distribution is a slightly biased version of the original distribution. Past mining depleted, on average, thicker seams which probably accounts for the relatively poorer fit in the upper tail of the estimated distribution as noted below. Alternatively, the distribution can be viewed as approximating today's distribution.

27. The total underground reserves in seams 28 inches or more in the measured and indicated categories is given by the Bureau of Mines as 2.2 billion tons. This is larger than the reserves of the coal-producing states of Alabama, Virginia, Tennessee, and Maryland. It represents 73 percent of the reserves of Eastern Kentucky. United States Bureau of Mines, Information Circular, 8655, "The Reserve Base of Bituminous Coal and Anthracite for Underground Mining in the Eastern United States."

28. The data classify reserves into the following sulfur intervals, where the limits are percent sulfur by weight:

$< .44$
.45 – .64
.65 – .84
.85 – 1.04
1.05 – 1.44
1.45 – 1.84
1.85 – 2.24
2.25 – 2.64
2.65 – 3.04
3.05 +

29 The reason for the behavior in the tail for the Montana data is that data were not recorded for higher overburden ratios. An arbitrary cut-off on overburden ratios was used. The cut-off overburden ratio decreased as depth of seam increased, so that the upper tail of the overburden distribution was truncated. So, the approximation is better than the data show. See (7).

30. See U.S. Bureau of Mines, "Strippable Reserves of Bituminous Coal and Lignite in the United States," Information Circular 8531, 1971.

31. Data on new deep mines come from the *Keystone Coal Manual* for 1973, 1974, 1975. The Western strip mine data is from the U.S. Bureau of Mines, *Projects to Expand Fuel Sources in Western States*, IC 8719, 1976. The Midwestern strip data combine observations from *Coal Age*, February 1977, State of Illinois, Department of Mines and Minerals, *Annual Report*, 1975, and *Keystone Coal Manual*, 1975.

32. Recent work of the Federal Energy Administration (17) does attempt to deal with depletion. The essence of their procedure is the assumption that all mines produce at a constant rate of output for twenty years. Depletion occurs as low-cost mines are exhausted at the end of twenty years. While certainly an improvement over earlier efforts, this assumption leads to anomalous results. For example, since the bulk of the expansion in low-sulfur production has taken place in the last five years, no low-sulfur mines are exhausted until 1990. The effect of cumulative output is zero for all low-sulfur coal in all regions until 1990. In that year, if output rates were to remain constant, there would be a large discontinuous jump in cost. This problem is contradicted by the recent rapid increases in low-sulfur coal prices, and differs substantially from the results of this paper.

REFERENCES

1. American Institute of Mining Engineers, *Mining Engineering Handbook*, Summer 1973.
2. Aitchison and Brown, *The Lognormal Distribution*, Cambridge University Press, 1957.
3. Allais, M., "Method of Appraising Economic Prospects of Mining Exploration Over Large Territories," *Management Science* (July 1957), 285–345.

4. Herfindahl, Orris, and C. Allen V. Knesse, *Economic Theory of Natural Resources*, Charles E. Merrill Publishing Company, Columbus, Ohi, 1974.
5. Hotelling, Harold, "The Economics of Exhaustible Resources," *The Journal of Political Economy*, Vo. 39, No. 2 (April 1931), 137–175.
6. Illinois Department of Mines and Minerals, *1973 Annual Coal, Oil and Gas Report*.
7. Matson, Robert E., and John W. Blumer, *Quality and Reserves of Strippable Coal, Selected Deposits, Southeastern Montana*, Montana Bureau of Mines and Geology, Bulletin 91, December 1973.
8. Pfleider, Eugene P. ed., *Surface Mining*. American Institute of Mining Engineers.
9. Simon, J. A., and W. H. Smith, "An Evaluation of Illinois Coal Reserve Estimates," Proceedings of the Illinois Mining Institute, 1968.
10. Stefanko, Ramani and Ferko, *An Analysis of Strip Mining Methods and Equipment Selection*, Research and Development Report No. 61, Office of Coal Research, May 29, 1973.
11. United States Bureau of Mines, "Basic Estimated Capital Investment and Operating Costs for Underground Bituminous Coal Mines," Information Circulars 8632 and 8641, U.S. Government Printing Office, Washington, D.C. 1974.
12. United States Bureau of Mines, "Cost Analyses of Model Mines for Strip Mining of Coal in the United States," Information Circular 2535, U.S. Government Printing Office, Washington, D.C., 1972.
13. United States Bureau of Mines, "The Reserve Base of Bituminous Coal and Anthracite for Underground Mining in the Eastern United States," Information Circular 8655, U.S. Government Printing Office, Washington, D.C., November 1974.
14. United States Bureau of Mines, "The Reserve Base of U.S. Coals by Sulfur Content, Information Circulars 8680, U.S. Government Printing Office, Washington, D.C., 1975.
15. United States Federal Energy Administration, *PIES Coal Supply Curve Methodology*, January 1976.
16. Williams, Cyril H., Jr., "Planning, Financing and Installing a New Deep Mine in the Beckley Coal Bed," *Mining Congress Journal*, August 1974.
17. Woodruff, Seth, *Methods of Working Coal Mines, Vol. III*, Pergamon Press, Elmsford, N.Y., 1966.

THE RATE OF PETROLEUM EXPLORATION AND EXTRACTION

Russell S. Uhler, UNIVERSITY OF BRITISH COLUMBIA

I. INTRODUCTION

Studies in the economics of petroleum energy and other exhaustible resources can be broadly classified into two groups: (a) those that have focused on price and extraction dynamics under different conditions of market demand and structure, production costs, and the existence of substitutes,[1] and (b) those which have been primarily concerned with describing petroleum reservoir production and cost conditions.[2] A common feature of these studies is that they have examined the production and price dynamics of a fixed stock of known reserves.

It is of course well recognized by both of these groups that the stock of

Advances in the Economics of Energy and Resources, Vol. 2, pp. 93–118.
Copyright © 1979 by JAI Press, Inc.
All rights of reproduction in any form reserved.
ISBN: 0-89232-079-6

known reserves in the earth's crust is not a fixed amount but is continuously altered by new discoveries. This has led the first group in the direction of viewing the problem as one of extraction from an uncertain rather than a known stock.[3] Their objective is to determine the impact of uncertainty of new discoveries on price and extraction profiles under a variety of conditions.

The second group faces the following two tasks if it is to provide a unified theory of petroleum energy production: (a) since relatively little has been done on the economics of the reserve discovery process, the understanding of this process must be expanded and brought up to the level of the economics of extraction, and (b) the dichotomy between these two stages of the total production process must be unified.[4] It is one purpose of this paper to help in each of these tasks. A second purpose is to further examine the extraction stage of the production process.

In building a model of petroleum discovery we cannot rely on the same framework used in determining optimal extraction policies from known reserves. The fundamental difference, of course, is that we can observe the initial stock of reserves in one case, but we cannot in the other. The difficulty of an unobservable stock from which new reserves are discovered is overcome in this paper by building an observable state variable into the model which acts as a substitute for the unobservable resource stock. This approach is described in the next section of the paper.

A primary objective of the paper is to study the optimal path of exploration in a petroleum play in which geological information is accumulated and aids in reservoir discovery but where discoveries serve also to exhaust the stock of undiscovered reserves.[5] The tendency for the largest reservoirs to be found early in the play is also recognized. Parameter estimates are used to generate a particular solution for the optimal path of exploration in one region in Alberta, Canada, and this path is then compared with the actual industry path.

Following the discussion of the exploration stage of petroleum production, a model of extraction from known reservoirs is developed. A special effort is made to integrate the physical behavior of reservoirs into an economic model of extraction and thus help to fill the gap between economic and engineering approaches to the problem. We will see that the optimal extraction rate is usually, but not always, determined by pressure conditions within the reservoir. We will also consider the economics of investment in pressure maintenance facilities.

The effect of royalty and profits taxes on extraction rates and on exploration effort is also considered. We will see that under fairly typical conditions that changes in these taxes will not directly affect extraction rates but may indirectly affect them through their effect on investment in pressure maintenance facilities. Consideration of the effects of these taxes on exploration effort, through their effect on the implied price of reserves in the ground, will serve as a vehicle for helping to unify the exploration and extraction stages of production.

2. A MODEL OF EXPLORATION

Let us begin by first considering those factors which we can expect to be important in explaining the rate of discovery of petroleum reserves. If we were to look elsewhere to other types of resource production situations for suggestions, we would find that the resource production rate depended primarily upon the rate of current effort and on the size or quality of the remaining resource stock. For any given level of current effort, the higher the size or quality of the remaining stock the higher the rate of resource production.

A similar rationale can apply in explaining the discovery of petroleum reserves. Current effort and the size of the stock of undiscovered reserves should be important factors in explaining the discovery rate although the level of geological knowledge is also important. The greater the knowledge of where to find petroleum the higher will be the discovery rate for any given amount of effort and undiscovered reserves. Furthermore, for any given level of current search effort and geological knowledge, the larger the stock of undiscovered reserves the higher is their rate of discovery.

The difficulty with such a specification for purposes of this paper, however, is that it is not operational simply because the initial stock of undiscovered reserves is unobservable. We need an observable variable which can be used as a substitute for the undiscovered resource stock. Fortunately an alternative is available.

Current exploration effort has four primary effects: (a) it directly affects the rate at which reserves are discovered, (b) it reduces the stock of undiscovered reserves to the extent that it results in discoveries, (c) it addes to our geological knowledge, and (d) it, by definition, increases the stock of cumulative effort. Thus, it is clear that the stock of cumulative effort is inversely related to the level of undiscovered reserves and may act as a substitute for it in the specification of a production function. Moreover, it has the advantage that its initial value is an observable quantity. It has the further advantage that it also reflects the level of geological knowledge, a variable which we mentioned earlier as being important in explaining the discovery rate and one which is ordinarily difficult to quantify.

How then do we expect the level of cumulative exploration effort to affect the reserves discovery rate for given levels of current effort? In the early stages of the exploration of a play or geographical region, increases in cumulative effort should lead to such significant additions to geological

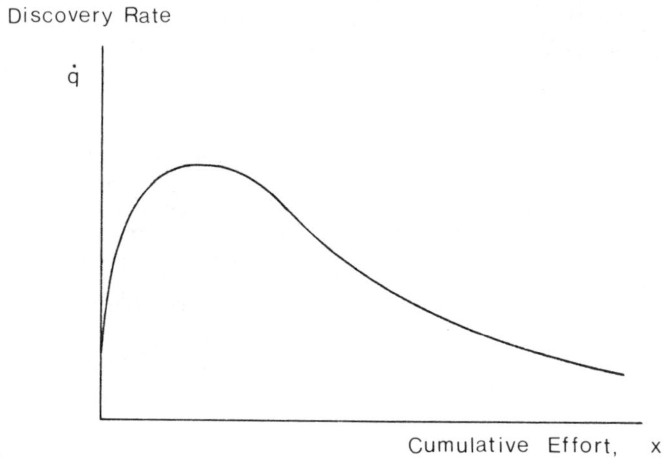

Figure 1.

knowledge that the discovery rate should rise even though the stock of undiscovered reserves is simultaneously being reduced. But because both the average size and number of undiscovered reservoirs begins to decline as cumulative effort grows, we would expect the impact of this effect to begin to dominate so that the discovery rate should begin to decline rather early in the exploration of the play. Thus the positive effect of geological knowledge on the discovery rate is soon dominated by the negative effect of the reduction in the stock of undiscovered reserves.

Figure 1 illustrates the relationship between the discovery rate, $\dot{q}$, and the stock of exploration effort, x, as described above. For initial values of x, the effect of geological knowledge dominates the effect of reduced reserves and the discovery rate rises with increases in x. But as x increases, the effect of reduced reserves begins to dominate the effect of additional geological knowledge and the discovery rate begins to fall. Thus we see that the marginal product of cumulative effort is at first positive but becomes negative after some critical value of x is reached.

To be more specific about the functional relationship described above and illustrated in Figure 1, let

$$dq/dt = \dot{q} = Q(x,v) = v^{\alpha}B\exp\left[-\beta(x-k)^{2}\right] \tag{2.1}$$

where $v = \dot{x}$, and A, α, β and k are parameters. This relationship states that for a given level of current effort, the reserves discovery rate increases until x reaches k, at which point it declines asymptotically toward the horizontal axis. The increasing part of the curve captures supremacy of the geological knowledge effect referred to earlier and the declining part represents the eventual domination of the declining amount of undiscovered reserves.

Equation (2.1) is, of course, only one possible representation of the reserves discovery process. It has the advantage that it is reasonably flexible since its centrality and spread change with variations in the parameters k and β. On the other hand, it tends to overstate the discovery rate for very low values of x since at $x = 0, \dot{q} > 0$. The true test of its usefulness as a general description of the discovery process, however, is an empirical matter and it has been shown to provide a good fit to the data from several regions in Alberta.[6]

In order to complete the technical conditions of the model we must now specify the supply conditions for exploration effort which are appropriate in this situation. As the input requirements of the exploration industry rise we would normally expect their supply price to either rise or remain relatively constant. In this paper both of these supply conditions are examined.

Specifically, the two supply conditions of exploration effort which are considered here are the case in which the supply price is a linear function of the rate of effort and the case in which it is a constant. If c(v) is the total cost of the exploratory effort rate v, then unit cost for the first case is given by

$$c(v)/v = a + bv \tag{2.2}$$

where a and b may be functions of time but are assumed to be constant here. The second case is where $b = 0$.

It should be emphasized that these are factor supply conditions which face the industry, or a social manager acting in its place, and not necessarily those which face the individual firm. Of course, no matter what the structure of the industry, if the industry supply price is constant so will be that price faced by the individual firms.

We now want to define the objective functional which is to be maximized with respect to the rate of exploration effort, v. Let us first consider the case in which the supply price of effort is rising. In this case a monopolist would simply want to maximize the discounted value of profits which is given by

$$J = \int_0^\infty [pQ(x,v) - c(v)]e^{-rt}dt \tag{2.3}$$

where r is the discount rate and p is the rice of reserves in the ground.[7] But if c(v)/v is upward sloping then J(x, v, t) is not the function which a social manager would want to maximize. Instead, he should maximize a functional which, in equilibrium, forces equality of the value of the marginal product of exploration effort and its supply price and not equality with the marginal factor cost. In the specific case in which the cost of effort is quadratic this is easily accomplished by maximizing

$$W = \int_0^\infty [pQ(x,v) - c^*(v)]e^{-rt}dt \tag{2.4}$$

where $c^*(v) = av + bv^2/2$.

In the case where the supply price of effort is constant and $b = 0$ the monopolist and the social manager both maximize the same functional. But whichever functional is maximized it is subject to the following constraints:

$$\dot{x} = v, x(0) = x_o. \tag{2.5}$$

Let us now maximize equation (2.4) subject to equation (2.5). To do this we form the Hamiltonian given by

$$H(v, x, \lambda, t) = [pQ(x, v) - c^*(v) + \lambda v]e^{-rt} \tag{2.6}$$

where $\lambda(t)$ is the costate variable. The costate variable may be interpreted as the effect of a unit change in the stock of effort on the maximum W; it may be thought of as the shadow price of cumulative effort. The Hamiltonian function itself has an economic interpretation. If it is multiplied by a small time interval, dt, we can see that it is the contribution to net social benefits of current effort during this time interval plus the contribution of the incremental adjustment to the stock of effort valued at its shadow price. naturally, our goal is to maximize this function with respect to the control variable, v, for every point of time. From the maximum principle the necessary conditions for optimality are given by

$$\partial H/\partial v = p\partial Q(x, v)/\partial v - \partial c^*/\partial v + \lambda = 0 \tag{2.7}$$

$$\dot{\lambda} = \lambda r - p\partial Q(x, v)/\partial x. \tag{2.8}$$

Assuming that equation (2.7) can be solved uniquely for v, given initial values of x and λ these differential equations trace out the entire time paths of v, x, and λ. Furthermore, we know that if a saddle point equilibrium solution exists, there is a set of initial values of x and λ which produce an optimal trajectory which passes through the equilibrium point. We will consider the characteristics of the optimal path of v for a given set of parameter values later in the paper.

Let us now consider the optimization conditions given by equations (2.7) and (2.8) with the restriction that $\alpha = 1$. In this case we now write

$$Q(x, v) = Q(x)v. \tag{2.9}$$

This restriction will remain in force throughout the rest of the paper and an empirical justification for it will be supplied later.

Equation (2.7) says that every instant optimality requires that v be chosen such that the value of the marginal product plus the shadow price of cumulative effort should equal the factor supply price. If v is positive, then it remains so, and x grows until its equilibrium level is reached when $\lambda(t) = 0$. At this point v is also zero and $pQ(x) = a$. Figure 2a illustrates this situation. Starting at zero, x grows until its equilibrium value x^* is obtained. In Figure 2a only the equilibrium marginal social cost function is shown. As x grows

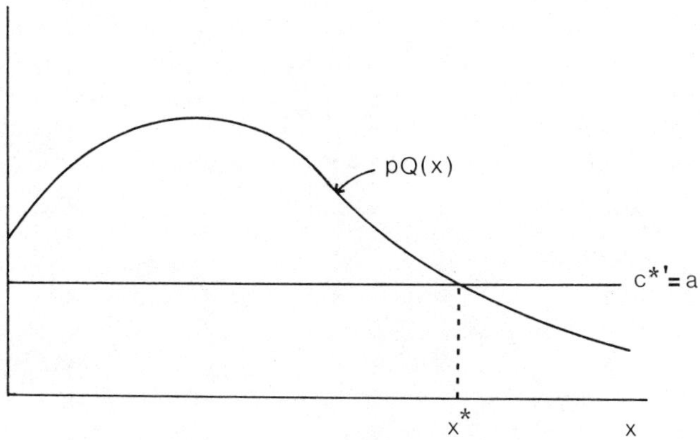

Figure 2a.

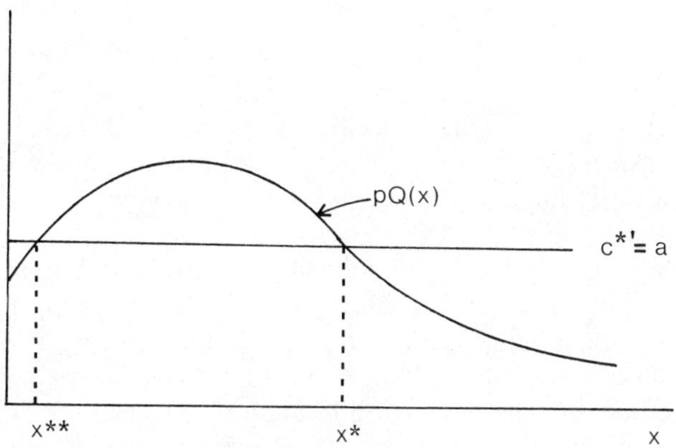

Figure 2b.

toward x*, v is not constant, so there is a series of such cost curves, one for each level of v. The difference in the two curves is $pQ(x) - c^{*\prime}(v) = -\lambda(t)$ so $\lambda(t)$ will typically be negative until x* is reached.

Given the symmetry of Q(x) it is apparent that there are two equilibrium points in our model. Although one of these points would probably not occur for positive x and thus would be of no significance in this problem, the parameter values of the problem may be such that both points of equilibrium would occur for positive x. Figure 2b illustrates this case. The study of the motions in the phase plane of this case make it apparent, however, that x** is an unstable equilibrium. Thus for an $x \lesssim x^{**}$ the trajectories will always diverge from this equilibrium. It does not mean, however, that in the event that such a situation arises that exploration would not occur. Although negative net benefits would be incurred in the early part of the exploration program, it may be worthwhile to sustain them in order to get positive

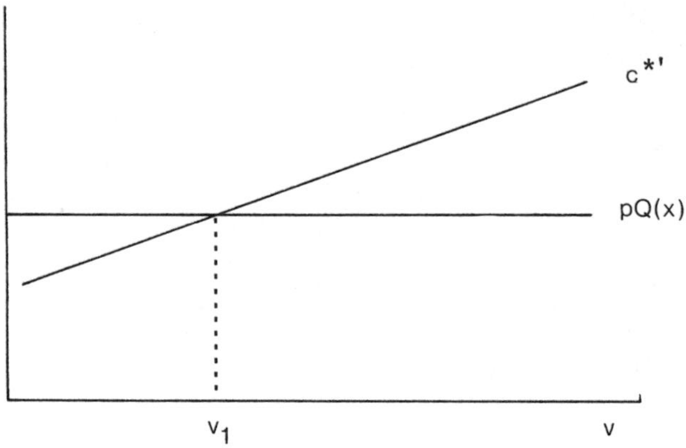

Figure 3.

benefits later. Moreover, even though an optimal time path for exploration effort would not exist until $x > x^{**}$, one would exist thereafter.

It is of interest to compare the exploration dynamics described above with a classical representation of the problem in terms of factor demand and supply. The demand function, of course, is just the value of the marginal product of v and for $\alpha = 1$ is pQ(x). The factor supply curve in this case is $c^{*\prime}(v) = a + bv$. Equating these and solving for v gives the classical determination of the optimal rate of effort. This is illustrated in Figure 3. Since $\alpha = 1$ the demand curve is horizontal, but since it depends on x it is shifting continuously so long as $v > 0$. The expected classical path of v then might be something like that illustrated in Figure 3. The factor demand and supply curves initially intersect at, say, v_1. This causes x to increase, which has the initial effect of shifting demand upward so that the optimal v increases. This upward shift continues over the range in which the marginal product of x is positive. But as we have already seen, eventually the marginal product of x becomes negative as the reservoir exhaustion and size effects begin to dominate and the demand curve shifts downward so that the optimal level of v declines toward zero. Thus the classical pattern of the optimal v is completely controlled by Q(x). Exploration effort will increase until Q(x) is maximized, after which it declines steadily with further increases in x.

There is one additional situation which we should consider since it is the classical analogy of the situation illustrated in Figure 2b. It may be that initially the demand and supply curves do not intersect at a positive level of v so that the optimal classical policy is apparently to do nothing. But if for some reason exploration were to begin, the demand curve would shift upward and an optimal positive v would then exist. This is analogous to the situation in Figure 2b where no optimal path of v existed so long as $x < x^{**}$.

Although these classical considerations are of interest, we already know that following the classical policy of equating factor demand and supply is not optimal. This is of course apparent from equation (2.7). The dynamic multiplier $\lambda(t)$ is always non-zero and has its own optimal path, as we have already noted. If this path is such that it is predominantly negative, as it is in the empirical example which we will consider later, then the classical policy would lead to a choice of v which was usually too large. The rationale is straightforward. If the marginal value of cumulative effort is negative then we would not want to accumulate it as fast as if its marginal value were zero as in the classical case. Thus the classical policy recommends a level of exploration effort which is too large. On the other hand, if the optimal $\lambda(t)$ is positive then the classical policy gives a level of effort which is too small.

Another feature of the model where $\alpha = 1$ is that it can be used to estimate the total discoverable and undiscovered reserves in the region or play.[8] To see this we note that

$$dq/dt = Q(x)v = Q(x)dx/dt. \tag{2.10}$$

This implies that

$$dq/dx = Q(x), \tag{2.11}$$

so that

$$q(\infty) = \int_0^\infty Q(x)dx \tag{2.12}$$

where q (∞) is total discoverable reserves. Total undiscovered reserves is simply q (∞) – q (x). Since we can estimate the parameters of Q(x) from the discovery data, estimates of the remaining undiscovered reserves can be made.

It will be noted, however, that such estimates are estimates in the tradition of M. K. Hubbert (6) in that they do not give adequate importance to economic factors. But this weakness can be overcome by a simple alteration of of our model. We can see immediately that the remaining undiscovered reserves that are of economic significance given current prices and costs are given by

$$q(x^*) - q(x) = \int_0^{x^*} Q(x)dx - \int_0^{x} Q(x)dx = \int_x^{x^*} Q(x)dx. \tag{2.13}$$

Since x* depends upon price and cost conditions, as is clear from the equilibrium equation $pQ(x^*) = a$, we can determine the impact of changes in these conditions on remaining undiscovered reserves using equation (2.13). For example, a rise in p or a reduction in *a* will raise the equilibrium x* and thus raise the level of undiscovered reserves which are of economic significance.

The Linear Case

In the previous section the model was nonlinear because of the nonlinearity of the function describing the cost of exploratory effort. In this section this function is assumed to be linear, in which case, combined with our restriction that $\alpha = 1$, the model becomes linear in v.

The objective functional is now expressed as

$$J = \int_0^\infty (pQ(x) - a)\, v e^{-rt}\, dt. \tag{2.14}$$

We want to maximize equation (2.14) subject to the same constraints as before, but we must now place the following restriction on the control variable

$$0 \leq v \leq v_{max} \tag{2.15}$$

where v_{max} is the maximum exploration capacity of the industry.

We want to consider three features of the linear model. They are: (a) the nature of the equilibrium x and how it compares with the nonlinear case, (b) the effect of known changes in price through time, and (c) the behavior of the optimal v.

The first two of these features are immediately apparent from the Euler equation when price is assumed to be an explicit function of time. From this equation we have

$$p(t)Q(x^*) - \dot{p}(t)Q(x^*)/r = a, \tag{2.16}$$

so that if $\dot{p}(t) = 0$ the equilibrium x is the same as before. However, if the industry or social manager can look forward to, for example, increasing prices over time, two opposing effects are apparent. A rise in p(t) causes the marginal product of v to rise, thus tending to increase the equilibrium value of x. But the second term in equation (2.16) suggests that to the extent that prices are expected to rise in the future the equilibrium x is reduced. This is an effect that we would not have detected from ordinary comparative dynamics.

Unlike the nonlinear case which we will consider in an example in the next section, the path of v in the linear case is easy to determine. So long as the marginal product of v exceeds its constant marginal cost, $v = v_{max}$, and x grows at this rate toward its equilibrium value. At $x = x^*$, $v = 0$. Thus we have "bang-bang" control which is usually associated with linear models.

The Parameter Values and the Characteristics of the Solution

In this section we want to consider the characteristics of the optimal path of exploration effort for a set of estimated parameters and for the case in which the unit cost of effort is a linear function of the rate of effort. Once the nature of the optimal path is determined it will be compared with the observed path. The difficulties of making such a comparison are also discussed. What we have been calling exploration effort will be measured by exploration drilling footage per unit of time.

For purposes of this example a relatively small region of Alberta was selected for analysis. The oil reservoir discoveries in the region are mostly part of a single petroleum play, but there have, of course, been discoveries in other plays. Thus the conditions generating the discovery data are roughly similar to those conditions under which the model was constructed.

Of particular interest is the estimate of the parameter α since its value thus far has been assumed to be unity.[9] The point estimate of α in this region is 1.05, but the variability of the estimate is such that unity is well within the probable range of the true parameter value. Furthermore, other regions in Alberta have provided estimates of α which are also close to unity and statistically very stable. Thus the assumption, $\alpha = 1$, seems to be reasonable and, moreover, considerably simplifies the mathematical aspects of the problem.

The only other parameter of the $Q(x, v)$ function which is of special interest to us here is k. It turns out that for this region the estimate of k is very small in magnitude, which means that over nearly the whole range of its value the marginal product of x is negative. In the classical framework this means that the demand for exploratory effort shifts downward nearly from the beginning of exploration activity. The reason for this empirical result is that in this region the largest reservoirs were discovered very early and their size simply swamped the effect of early growth in the number of reservoirs which were discovered as cumulative effort grew.

The determination of the appropriate price and cost parameters is particularly difficult. We have already seen that variations in these parameters change both x^* and the optimal path of v. In reality, these parameters are changing continuously due to changing technical and economic conditions which are exogenous to our model. Thus any comparison between the optimal path of v, based on fixed parameters or parameters which are expected to change in a certain way over time, and the observed industry path which is necessarily generated under the more complex set of conditions must be interpreted with care.

With these difficulties in mind let us consider the determination of the cost of effort function parameters. As usual, in order to determine these parameter values, they must be estimated from the historical data. But other factors besides the rate at which it is supplied also affects the supply price of drilling footage (effort). The factors of particular significance are drilling technology and the depth of wells. To hold the effects of changes in drilling technology constant, cost data for relatively short periods of time were analyzed. The unit cost of exploratory drilling was regressed on its rate and the depth per well. The depth variable was included to eliminate the effects of variation in its value so that the independent effect of effort on its unit cost could be determined. This was done for two five-year periods with observations based on sixty-day intervals. Both periods showed significant (at the .10 level) coefficients on the effort variable suggesting that the linear unit cost function is a reasonable approximation. However, as one would have expected, the magnitude of the coefficients differ significantly between the two periods.

In view of these results and the fact that price also changed during and between the two periods, the optimal path of v was determined for both sets of parameter values with the price being the average in each period. Each set produces a similar path of the optimal v. In fact, over a fairly wide range of values of the price and cost parameters the general pattern of the optimal v is the same. It is this fact that is most important to us, since it suggests that the general pattern of the optimal v is insensitive to upward or downward shifts in the parameter values which we know have occurred but which cannot be explicitly recognized in the solution.

The solid line in Figure 4 illustrates a typical optimal path of v as a function of x. Such a path is produced by solving the differential equation system (2.7) and (2.8) using those initial values of x(0) and $\lambda(0)$ which produced a trajectory passing through the equilibrium x. Since it is easily shown that x* is a saddle point equilibrium, we know there is only one such optimal trajectory.

As we might have expected from our earlier discussion of the parameters

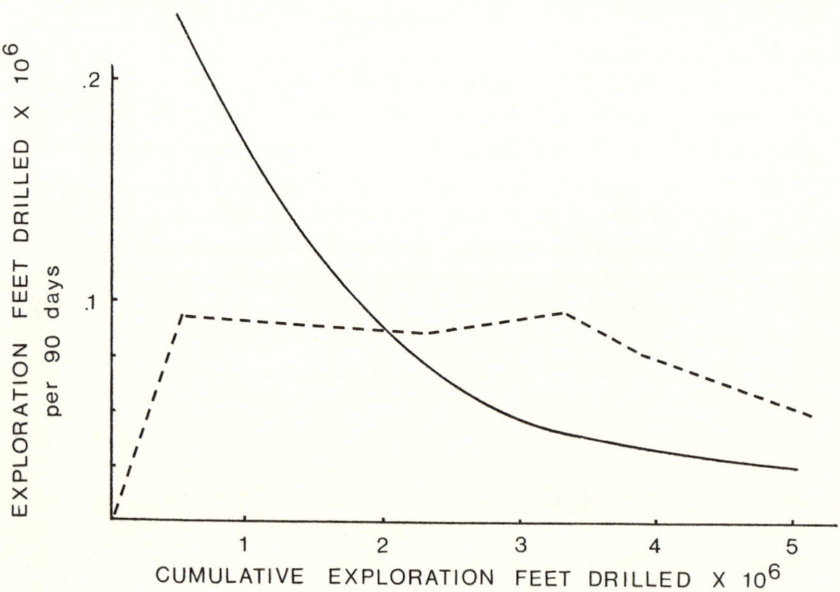

Figure 4.

of $Q(x)$, the optimal v starts out at a relatively high value and declines steadily with the growth of x. And as $x \rightarrow x^*$, $v \rightarrow 0$. The dashed line in Figure 4 shows the observed industry rate of exploratory effort in the region.

It is of interest to compare the two curves in Figure 4 and to speculate on the possible reasons for their difference. Except for the hump at around $x = 3.5$ the general character of the observed path is one of sharp increase followed by a slow decline. On the other hand, the optimal path is characterized by a relatively high level of exploration effort at the beginning, followed by a sharp decline, and passes below the observed path at about $x = 2.0$. The rate of decline in the optimal path then becomes much slower but it is always below the observed path. Thus, except for the very general pattern of decline in v with the growth in x which is exhibited by both paths, it is fair to say that they are quite different. Does this mean, then, that the industry path indicates an exploration pattern which is not socially optimal? Not necessarily, for as was mentioned earlier the results need to be interpreted with care.

Let us consider the most obvious difference between these two curves, that which occurs at the beginning of the exploration of the region. There are at least two good reasons why the observed and the optimal paths should differ at this point. They are: (a) the $Q(x, v)$ function is not well suited for describing the initial part of the relationship between the discovery rate and cumulative effort, and (b) our model assumes perfect certainty on the part of a social manager when, in fact, this is obviously not the case.

At the beginning of the exploration of a play, exploration effort will often result in no discoveries. Thus, as x grows in this range, $\dot{q}$ is zero. Suddenly a discovery is made which usually leads quickly to other discoveries. If these initial discoveries are large by the standards of the play then $\dot{q}$ suddenly jumps to a large value and may even decline from that point on. Clearly our $Q(x, v)$ function is not well suited for describing this kind of occurrence even though, over the whole range, it may provide a fairly good description. Thus we would not expect the optimal path to show any increase in v with increases in x under these circumstances. Only if the parameter k were relatively large would the ability of $Q(x, v)$ to describe discoveries in the initial range of x be improved. But even if we disregard the initial fast growth in discoveries along the observed path, the highest level that the observed exploration rate ever reaches is well below the highest optimal level.

Some of this difference is undoubtedly due to the effect of uncertainty. After all, the industry begins exploration in a state of high ignorance whereas, in our model, the social manager has perfect knowledge of the discovery rate as a function of current and cumulative effort and the parameter values. If the effect of uncertainty is generally to reduce the level of investment in an activity, then one would probably expect the observed path in this example to be below the optimal path.

If the industry behaves in a socially optimal way, as information accumulates and uncertainty is reduced we might expect the observed and optimal paths to get closer as x increases. Indeed, this seems to occur as x grows beyond $x = 3.5$ with the industry path now always above the optimal one. Some of the difference in this range, however, is undoubtedly caused by exploration activity which is directed toward investigating new or possible new plays in the region. Had we been able to identify exploration effort which was clearly directed toward one play, the observed and optimal paths would probably have been closer. Thus we may conclude that, all things considered, the observed pattern of exploration activity does not diverge so much from the optimal pattern as is first suggested by Figure 4.

3. EXTRACTION FROM PETROLEUM RESERVOIRS

The second stage of the petroleum production process involves the extraction of petroleum from known reservoirs which were discovered in the first stage. The primary economic link between the two stages is the value of reserves in the ground, which is determined within the extraction stage of production. Thus, in order to determine the effects of a variety of economic conditions and policies on exploration activity it is essential to understand their effect on the extraction stage of the production process. Moreover, the effect of these conditions and policies on the rate of extraction is of interest in its own right.

In this section we want to examine the economic and technical factors which determine the rate of extraction from petroleum reservoirs. Special attention is given to the task of integrating reservoir pressure dynamics into an economic model of extraction. One can hardly neglect reservoir pressure conditions since this pressure is, after all, the force which drives the oil to the earth's surface once a well has been completed to a producing formation. Without it production simply cannot occur. Not only does such an integration seem to model the actual situation more closely, it has the additional advantage of allowing us to focus more closely on the economics of investment in pressure maintenance programs. Such programs are capable of producing significant increases in the rate of extraction and in recoverable reserves.

Some Basic Properties of Petroleum Reservoirs

Before introducing reservoir pressure dynamics explicitly into a model of petroleum extraction, it is useful to summarize some of the important properties of petroleum reservoirs which will be of use to us here. The petroleum engineering literature sets forth a number of wellbore pressure equations which are solutions to the basic differential equations which are said to represent the physical characteristics of a reservoir system. A number of facts about the pressure equations are noteworthy. First of all, the functional forms which are applicable to any situation are dependent upon assumptions about the size and shape of the reservoir and its boundary conditions. The boundary conditions are particularly important since they indicate the nature of the so-called drive mechanism. For example, if the reservoir is bounded by impermeable rock so that no fluid can enter the flow system, the pressure in the reservoir decreases with time as a result of the depletion of mass in the system. Such a system is said to be one of several types of gas-drive systems. On the other hand, if the outer boundary of the reservoir is in contact with water and the water can move into the pore volume of the rock as oil and gas are produced, then the pressure decline through time may be very small. Even though individual wells may go out of production as water displaces oil and gas, the pressure in producing wells remains nearly constant. Such a drive mechanism is called water drive. Pressure equations have been developed to account for each of these boundary conditions.

But petroleum reservoirs are individualistic and many show a combination of water and gas drive, so that rarely do actual conditions fit the idealized ones on which the pressure equations are based. Nevertheless, we can make use of them to achieve an adequate degree of approximation and to understand some general principles. To illustrate further, consider the general wellbore pressure equation for a variable extraction rate, $z_w(t)$, which is given by[10]

$$P_w(t) = P_w(0) - \theta\left[z_w(0)f(t) + \int_0^t \dot{z}_w(t)f(t-\tau)d\tau \right] \tag{3.1}$$

where $P_w(t)$ is the well pressure at time t, f(t) is a function of time and some geological parameters, and θ is a constant which is inversely proportional to reservoir size. We can immediately see from this equation that pressure at time t depends upon the previous time path of the production rate and on the nature of the function f(t). This equation holds for a number of reservoir boundary conditions since they are essentially differentiated by the specification of f(t).

It is useful to consider the following simple specification of f(t). It is, as we shall see a specification which yields a linear pressure decline for a bounded reservoir. Although linear pressure decline is not normally expected in bounded reservoirs until production has occurred for some time, it will serve as an adequate approximation for our purposes. Moreover, this formulation allows us to introduce the approximate effects of an unbounded system not directly through the specification of f(t), but by viewing it as a costless pressure maintenance system. Thus we let

$$f(t) = t + a_1$$

where a_1 is a constant. This yields

$$P_w(t) = P_w(0) - \theta\left[\int_0^t z_w(\tau)d\tau + a_1 z_w(t)\right]. \tag{3.2}$$

This pressure equation is probably more intuitive than the previous one since it shows that current pressure depends inversely on cumulative production. It is also clear, however, that for any given level of cumulative production the current pressure is inversely related to the production rate. This means that if a well is produced very quickly to achieve a given level of cumulative output, the pressure will be lower than if the same cumulative output had been produced over a longer period of time. Since well pressure constrains the feasible output rate, in the absence of pressure maintenance this could mean a reduced total amount recovered. Thus, there is an apparent trade-off between the rapidity of producing a given total and the total amount which can be produced.

Another noteworthy aspect of equation (3.2) is that reservoir pressure depends on the *ratio* of cumulative output to reservoir size, since θ is inversely proportional to reservoir size. Thus, to the extent that the potential reservoir extraction rate depends on reservoir pressure it will also depend on the ratio of the remaining reserves in the reservoir to initial reserves. This serves as a justification for using remaining reserves as an argument in petroleum production and cost functions. It should be noted, however, that this applies to production from an individual reservoir and that the use of remaining reserves in specifying an aggregate relationship should be approached with care.

The Basic Extraction Model

The discussion of wellbore pressure dynamics given above was for a single well in an oil reservoir. To make use of these results in the following analysis I will assume that reservoirs are homogeneous, so that pressure in each well at every point in time is the same and thus equal to the reservoir pressure. In this setting the pressure dynamics for a single well in a reservoir can be used to describe the behavior of the entire reservoir. We simply substitute reservoir pressure and production rates in place of their well counterparts.

We will also use the simplified version of the f(t) function discussed earlier by setting $a_1 = 0$. Thus the reservoir pressure will depend only on cumulative production.

Let us begin by assuming that firms producing oil reservoirs act as price-takers and that if parts of a reservoir are owned by different firms it is managed as a unit. This avoids common property difficulties and is, in fact, representative of current reservoir management practice.

We will assume that the objective is to maximize a profit functional given by

$$L = \int_0^T [\rho(t)z(t) - c(z(t))]e^{-rt}dt \tag{3.3}$$

where z(t) is the *reservoir* production rate, $\rho(t)$ is the current wellhead price of oil, c(z(t)) is the cost function, r is an appropriate interest rate, and T is the terminal time which is chosen freely.

The usual resource constraint to which this maximization is subject is given by

$$\dot{S}(t) = -z(t), \qquad S(O) \geq 0, S(T) \geq 0 \tag{3.4}$$

where S(t) is the stock of oil in the reservoir. The initial stock, S(0), is interpreted as the maximum physical amount contained in the reservoir. The amount of it which is ultimately recovered depends largely on economic factors.

Up to this point the model is identical to the usual simple model of the extraction of an exhaustible resource under conditions of certainty. But in the production of petroleum reservoirs we must recognize another, less tangible, kind of resource which is crucial to the oil-production process. This is of course reservoir pressure, since it is this pressure that drives the oil to the earth's surface. Thus it is necessary to introduce a pressure constraint into the model. Using equation (3.2) with $a_1 = 0$ this constraint is given by

$$\dot{P}_w(t) = \dot{P}(t) = -\theta z(t) \tag{3.5}$$

where $\dot{P}(t)$ is the time rate of change of reservoir pressure which, by assumption, is the same as $\dot{P}_w(t)$.

Finally we must establish the relationship between reservoir pressure and its allowable production rate. It seems reasonable to suppose that

$$z(t) \leq h(P(t)) \tag{3.6}$$

where $h'(P(t)) > 0$. Moreover, we can reasonably assume that unit increases in P(t) lead to smaller than unit increases in allowable production so that $h''(P(t)) < 0$.

The objective is to maximize equation (3.3) subject to equations (3.4), (3.5) and (3.6). To do this we form

$$\begin{aligned} H = \{\rho(t)z(t) - c(z(t)) + \lambda_1(-z(t)) + \lambda_2(-\theta z(t)) \\ + \mu[z(t) - h(P(t))]\}e^{-rt} \end{aligned} \tag{3.7}$$

where λ_1 and λ_2 are dynamic multipliers and

$$\mu \begin{cases} = 0 & z(t) < h(P(t)) \\ < 0 & z(t) = h(P(t)). \end{cases} \tag{3.8}$$

Maximizing H with respect to z(t) leads to

$$\rho(t) - c'(z(t)) - \lambda_1 - \theta\lambda_2 + \mu = 0 \tag{3.9}$$

$$\lambda_1 = \lambda_1 r \tag{3.10}$$

$$\dot{\lambda}_2 = \lambda_2 r + \mu h'(P(t)). \tag{3.11}$$

Let us assume that pressure declines naturally with cumulative production as given by equation (3.5) and that the resource stock constraint is not binding, i.e., pressure runs out before the reservoir is exhausted. We also assume initially that $c'(z) > 0$ but that both ρ and $c'(z)$ are invariant over time.

Let us interpret the equilibrium conditions given by equations (3.9) – (3.11). It is well known that so long as $c'(z) > 0$ the optimal rate of production from a known resource stock is below the $\rho = c'(z)$ rate. Equation (3.9) recognizes this by way of the positive multipliers $\lambda_1(t)$ and $\lambda_2(t)$. In equilibrium they both must grow exponentially at rate r except when the production-pressure constraint is binding and the growth of $\lambda_2(t)$ is modified in recognition of this condition.

Several possible paths of the optimal production rate arise depending on the nature of the marginal cost and h (P) functions. If the marginal cost function rises sharply then the optimal production rate will begin at a relatively low level and decline slowly over time since its rate of decline is easily shown to be inversely proportional to $c''(z)$. In this case we might naturally expect that such low production rates would not encounter the pressure constraint initially and may never do so.

A much more likely situation is where marginal costs increase very slowly with the production rate. In this case we would expect the desired production rate to be high enough so that the pressure constraint is immediately encountered so that the production rate would be governed by the differential equation

$$\dot{z}(t) = -\theta h'(P)z(t) \tag{3.12}$$

and would thus exhibit exponential decline at rate $\theta h'(P)$.

Another interesting and widely assumed case is where $c''(z) = 0$. Since the Hamiltonian as linear in z(t) the first order conditions will not contain it. However, it is clear that the optimal policy is to choose z(t) as large as possible since both ρ and $c'(z)$ are invariant over time. In many problems of this type the upper bound on z(t) is not always immediately apparent, but in this problem we already know that equation (3.6) provides such an upper bound. The constraint is encountered immediately and governs the whole production history of the reservoir. Thus in the case of constant marginal cost we would expect exponential decline of the production rate over the entire production period of the reservoir.

We have focused on the qualitative behavior of the optimal rate of reservoir production under varying reservoir cost conditions. Only in the case of constant marginal costs can we be certain that exponential decline will describe the optimal path of z(t) over the whole production period of a reservoir. Other cost situations will likely provide a mixture of exponential and other patterns of decline since we cannot always expect the optimal choice of z(t) to result in $z(t) = h(P)$. We do know, however, that the flatter the marginal cost function is, the more likely the pressure constraint is to be binding over a large portion of the production history of the reservoir.

It is useful now to shift our focus from the path of the extraction rate to the qualitative behavior of the optimal path of $\lambda_2(t)$. We are particularly interested in $\lambda_2(t)$ since it can be interpreted as the implied marginal value of reservoir pressure and thus has obvious importance to any discussion of the economics of investment in pressure maintenance systems.

Let us consider a case where cost conditions and wellhead price require that the desired production rate encounter the pressure constraint immediately. Certainly constant or near constant marginal cost conditions provide such an example. What then is the qualitative nature of the optimal path of $\lambda_2(t)$? Eliminating $\mu(t)$ between equations (3.9) and (3.11) under the conditions that $\lambda_2(t) = 0$ gives the following isocline

$$\lambda_2(t) = h'(P)[\rho - c'(h(P)) - \lambda_1]/(r + \theta h'(P)). \tag{3.13}$$

It is easily shown that $\left.\frac{\partial \lambda_2}{\partial P}\right|_{\dot{\lambda}_2 = 0} < 0$ so long as $h''(P) < 0$. Thus we can be certain that the optimal path of $\lambda_2(t)$ increases as P(t) declines, and since P(t) declines over time we therefore know that $\lambda_2(t)$ increases over time. This is true even when $r = 0$. Although this rise in the marginal value of pressure is not especially significant so long as we do not allow the artificial control of reservoir pressure, when the model is extended to allow such control its importance is immediately apparent.

Pressure Maintenance

In the previous section, oil was produced by the natural pressure system of the reservoir. It was noted that one of the dynamic multipliers could be conveniently interpreted as the implied marginal value of reservoir pressure. Moreover, an effort was made to determine the nature of the time path of this multiplier so that it might reveal some aspects of the economics of investment in pressure maintenance. In this section we want to pursue this investigation further.

If we are to allow pressure to be maintained artificially by appropriate expenditures we must modify some of the conditions of our model. The most obvious modification is to the dynamic equation (3.5) which now becomes

$$\dot{P}(t) = \theta(m(t) - z(t)) \tag{3.14}$$

where m(t) is simply the rate of fluid injection into the reservoir. It is also convenient to separate the cost components associated with the extraction and injection systems so that the cost function can now be written as

$$c(z(t), m(t)) = c_1(z(t)) + c_2(m(t)). \tag{3.15}$$

The problem now has two control variables, the extraction rate z(t), and the injection rate, m(t), with respect to which we want to maximize the new profit functional incorporating the new cost equation (3.15). To do this we form the Hamiltonian given by

$$H = \{\rho z - c_1(z) - c_2(m) + \lambda_1(-z) + \lambda_2(-\theta z + \theta m) + \mu(z - h(P))\}e^{-rt} \tag{3.16}$$

where $\mu(t)$ is defined as before. Maximizing with respect to both z(t) and m(t) we get equations (3.10) and (3.11) plus

$$\rho - c_1'(z) - \lambda_1 - \theta\lambda_2 + \mu = 0 \tag{3.17}$$

$$-c_2'(m) + \theta\lambda_2 = 0. \tag{3.18}$$

It is clear from equation (3.18) that the injection rate should be set so that at every point in time the implied marginal value of pressure should equal its marginal cost.

We are now in a position to consider a possible path for the rate of fluid injection. Consider the case in which the pressure constraint is initially binding on the choice of the optimal z(t). We have seen that in this case $\lambda_2(t)$ assumes an initial positive value and increases over time. The situation is illustrated using a diagram such as that in Figure 5.

In this illustration $\lambda_2(t) < c_2'(m)$ initially so that $m(t) = 0$. But $\lambda_2(t)$ will rise in value, and when it rises to a level so that it equals marginal injection costs for some $m(t) > 0$, then it will be worthwhile to begin pressure maintenance. Of course, it may be that $\lambda_2(t)$ will not rise to a level where $m(t) > 0$ before either pressure or the resource stock is exhausted.[11]

Once it becomes worthwhile to begin pressure maintenance the path of z(t) is no longer governed by equation (3.12) but by

$$\dot{z}(t) = \theta h'(P)(m(t) - z(t)). \tag{3.19}$$

Reservoir pressure will continue to decline and its implied value to rise so long as $m(t) < z(t)$. If conditions are such that $m(t) \rightarrow z(t)$ then the optimal path of z(t) over the whole production period will look something like

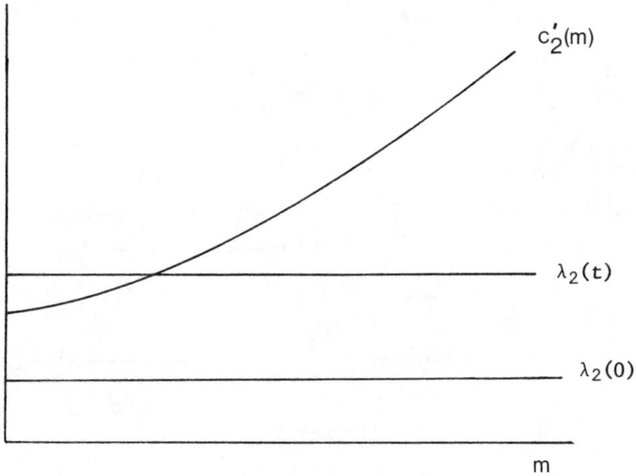

Figure 5.

Figure (6a). It begins with exponential decline up to the point where pressure maintenance begins after which its decline rate becomes smaller. The extraction rate becomes more nearly constant as $m(t) \rightarrow z(t)$.

This description of the optimal path of z(t) is one which is commonly observed in producing reservoirs. They often begin with a period of exponential-like decline followed by a period of nearly constant production once investment in pressure maintenance and/or pumping facilities has been made. This is one of the reasons why attempts to fit exponential functions to the whole production history of oil reservoirs has often been unsuccessful.

The path of m(t) in this example is something like that illustrated in Figure 6b. It is initially zero until it becomes worthwhile to begin pressure maintenance after which it rises towards a constant level at z(t). If it reaches this level then pressure remains constant until the reservoirs' recoverable reserves are exhausted. Of course, conditions may be such that pressure declines to a low enough level to halt extraction before recoverable reserves are exhausted.

Taxing Crude Oil Extraction

In recent years a variety of schemes for taxing the production of crude oil have been studied. They include: bonus bidding, which may be considered as a front-end tax prior to actual production; royalties; rental payments; oil pledges; profits taxes; and various mixtures of the above. Bonus bidding is recognized as best under the idealized conditions of competition and complete certainty. Royalties spread some risk but are said to lower the optimal production rate, cause early shutdown of producing reservoirs,

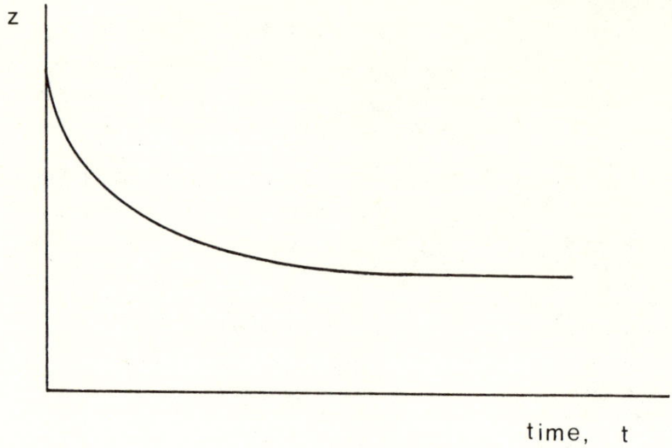

Figure 6a.

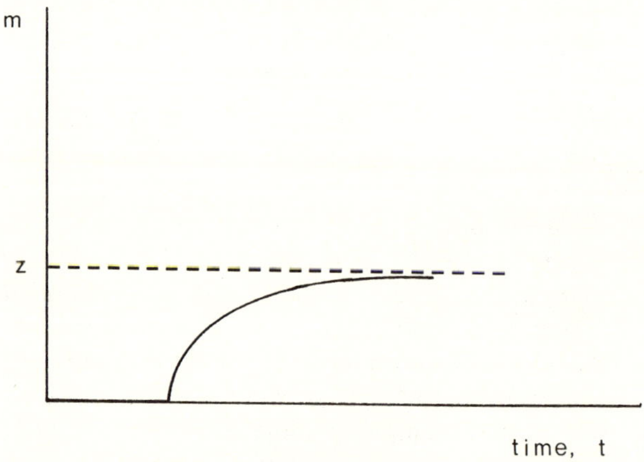

Figure 6b.

and reduce levels of exploration. The other schemes all have their own disadvantages except for profits taxes which are said not to impede exploration or to lower extraction rates or to lead to early shutdown. They also have significant risk-sharing advantages.

In this section I do not intend to provide a lengthy analysis of the effect of each of these taxation schemes in terms of the model given in this paper. What I want to do is simply consider the effect of royalty and profits taxes on crude oil extraction decisions and also use them to illustrate the link between the exploration and extraction stages of petroleum production.

In the simple fixed stock extraction model in which reservoir pressure conditions are not taken into account and where $c'(z) > 0$, the imposition of a royalty distorts the time path of extraction whereas a profits tax will not. But as we have already noted, cost conditions in petroleum extraction are usually such that the path of the optimal extraction rate is largely governed by reservoir pressure conditions. In this situation the imposition of a royalty will not directly affect the optimal extraction rate. To see this, note that when $c'(z) > 0$ the effect of a royalty is to lower the desired unconstrained extraction rate. But so long as its effect does not lower the desired rate below the $z(t) = h(P)$ level, the extraction path is still governed by the differential equation (3.12) so the royalty has no effect. If $c'(z) = \text{constant}$, then a royalty will never have a direct effect on the extraction rate so long as the net price exceeds the marginal cost. Since a tax on profits should not affect the extraction profile either, (in contrast with the usual comparison of the two types of taxes discussed above) the taxes are equivalent in this respect.

Suppose now that we consider the effects of these taxes when investment in pressure maintenance is allowed. Although neither are likely to have direct effects on the extraction rate, both will reduce the marginal value of pressure and thus lead to smaller investment in these facilities and to lower extraction rates below what would otherwise be the case. To see this, consider equation (3.13) which we shall modify to include the royalty tax rate, R, and the profits tax rate, γ. This equation is now written

$$\lambda_2(t) = h'(P)\{[\rho(1 - R) - c_1'(h(P)](1 - \gamma) - \lambda_1\}/(r + \theta h'(P)). \quad (3.20)$$

An increase in either R or γ will lower $\lambda_2(t)$ for all levels of P. Since the isocline is shifted downward as a result of these taxes, we would expect the optimal path of the marginal value of pressure, $\lambda_2(t)$ to shift downward at each point in time. This means that the path of m(t) also shifts downward so that pressure maintenance is less likely to begin at all and, if it does, at a lower level than would otherwise have been the case. The extraction rate is lowered and pressure is more likely to decline to zero with recoverable reserves still remaining. On the other hand, reduced taxes bring about pressure maintenance sooner and at higher levels, therefore reducing the likelihood that pressure conditions will affect the reserves recovery percentage. Thus, it is through their effect on the timing and rate of injection of fluids that both profits and royalty taxes effect extraction rates and recovery percentages. There is no reason to prefer one to the other in this respect.

In our model of the exploration stage of production the unit value of reserves in the ground, p, is the price to which exploration decision-makers respond. We saw that changes in this price change both the equilibrium level of cumulative effort and the path of current effort through time. Even though this price is not normally observed in the market, since most firms are engaged in both exploration and extraction, it is implied in the extraction stage of production by $\lambda_1(t)$ and serves as the primary economic link between the two stages. Increases in this price stimulates exploration incentives, and decreases retard them.

In the optimal depletion of individual reservoirs we saw that $\lambda_1(t)$ increases as both S(t) and P(t) decline, thus creating a price incentive for additional exploration as reserves are depleted. But it should be noted that the path of $\lambda_1(t)$ in our extraction model behaves as if no additional reserves will ever become available, i.e., that extraction is from a fixed stock of known reserves. We know, however, that exploration will turn up additional reserves so that in a fully integrated model of the two stages of production we would expect the growth in $\lambda_1(t)$ to be reduced, so that the effect of exploration in the integrated setting would be to reduce exploration incentives below what our more loosely connected model would predict.

Let us now turn our attention briefly to the effect of royalty and profits taxes on exploration incentives. We can see from equation (3.20) that increases in either royalty or profits taxes will reduce $\lambda_1(t)$ for all levels of P(t) (and S(t) as well) because of the nearly symmetrical relationship between $\lambda_1(t)$ and $\lambda_2(t)$. Thus, these taxes will affect exploration incentives through their effect on the price of reserves in the ground since we already know that changes in this price affects both the equilibrium level of cumulative exploration effort and its path through time. However, if profits are defined to include exploration costs then a tax on profits will presumably have no effect on exploration activity.

4. CONCLUDING REMARKS

Early in this paper we considered a model of the discovery of petroleum reserves in a single play or a region which is dominated by a single play. We then turned our attention to the rate of extraction from individual reservoirs and concluded that when marginal extraction costs increased slowly or not at all, with increases in the extraction rate, that the optimal path of the extraction rate would be governed by reservoir pressure dynamics. Moreoever, we concluded that extraction rates should exhibit exponential-like decline behavior appropriately modified by artificial or natural pressure maintenance. Since our discovery model explains the discovery of several reservoirs in a play, it seems natural to consider the behavior of the aggregate rate of extraction from all discovered reservoirs. Although a formal analysis of this aggregation problem is not intended here, a few conclusions are possible.

First, it is fairly obvious that the aggregate extraction rate would not look anything like that which is optimal for each individual reservoir. The reason, of course, is that the extraction from discovered reservoirs begins at different points in time. The behavior of the aggregate will depend on the extraction rates of the individual reservoirs and on their rate of discovery. We would normally expect that this aggregate would increase at first (even if all reservoirs declined exponentially) because of the growth in the discovery rate but that eventually the aggregate rate of extraction would fall as both the reservoir extraction and discovery rates declined. The precise determination of the optimal aggregate path using our model is complex. Indeed, to do it analytically we would need to make some rather special assumptions about the characteristics of the reservoirs.

Just as extraction rates aggregated to the level of the play will show slower rates of decline than will those for individual reservoirs, further aggregation over many plays or regions will likely decrease this rate of decline still further. Nevertheless, the fact that the resource is finite in amount in any given area suggests that ultimately a decline in the aggregate extraction rate will occur. This seems to be happening now for the United States as a whole, for example.

In this paper I have presented a model explaining the rate of discovery of individual reservoirs. The two stages of production were loosely connected via the price of reserves in the ground. The extraction model attempted to use some of the characteristics of reservoirs which are stressed by engineers as being important. Although we have achieved more detail in describing this stage of production than most economic models and have gained some interesting insights into factors affecting extraction paths and pressure maintenance decisions, it may be necessary to sacrifice some of this detail to achieve a fully integrated model which is analytically tractable. Only the future can tell.

FOOTNOTES

I would like to thank Colin W. Clark, Keizo Nagatani, and Hayne Leland for helpful discussions, and Keith Wales for computer programming. The work was supported by The Canada Council.

1. This group of studies tends to follow the pioneering work of Hotelling (5) of which Solow (10), Stigliz (11), and Nordhaus (9) are notable examples.
2. See Bradley (1) and Kuller and Cummings (7).
3. See Dasgupta and Stiglitz (3) and Gilbert (4).
4. Uhler (12) has recently provided a stochastic model of the discovery process from which estimates of marginal discovery costs were determined. Cox and Wright (2) provide a unified model of the discovery and extraction stages.
5. A petroleum play is defined as a search for petroleum in a geological horizon in which it is known to occur. Petroleum plays are usually associated with a particular geographical region.
6. See Uhler (12).
7. The price of reserves in the ground is the appropriate price to which the explorationist responds. It is determined directly as the difference between the wellhead price and average extraction costs evaluated at the optimal extraction level.
8. I am grateful to Colin W. Clark for pointing this out to me.
9. The estimation of the parameters of the Q(x, v) function will not be considered in detail here. For this detail, see Uhler (12).
10. A discussion of this equation may be found in Matthews and Russell (8).
11. Of course, with exponential decline pressure is never entirely exhausted, but it will become small enough to prevent further extraction from taking place.

REFERENCES

1. Bradley, P. G., *Economics of Crude Petroleum Production*. North-Holland, Amsterdam, 1967.
2. Cox, J. C., and Arthur W. Wright, "The Determinants of Investment in Petroleum Reserves and Their Implications for Public Policy," *The American Economic Review*, March 1976.
3. Dasgupta, P., and J. E. Stiglitz, "Uncertainty and the Rate of Extraction under Alternative Institutional Arrangements," Technical Report No. 179, Institute for Mathematical Studies in the Social Sciences, Stanford University, March 1976.
4. Gilbert, R., "Optimal Depletion of an Uncertain Stock," Technical Report No. 207, Institute for Mathematical Studies in the Social Sciences, Stanford University, May 1976.
5. Hotelling, H., "The Economics of Exhaustible Resources," *Journal of Political Economy*, April 1931.
6. Hubbert, M. K., *Energy Resources*, Publications 1000–D, Washington National Academy of Science—National Research Council, 1962.
7. Kuller, R. G., and R. G. Cummings, "An Economic Model of Production and Investment for Petroleum Reservoirs," *American Economic Review*, March 1974.
8. Matthews, C. S., and D. G. Russell, *Pressure Buildup and Flow Tests in Wells*, Society of Petroleum Engineers of AIME, New York, 1967.
9. Nordhaus, W., "The Allocation of Energy Resources," *Brookings Papers on Economic Activity*, 3, 1973.
10. Solow, R. M., "The Economics of Resources or the Resources of Economics," *American Economic Review*, May 1974.
11. Stiglitz, J. E., "Monopoly and the Rate of Extraction of Exhaustible Resources," Technical Report No. 169, Institute for Mathematical Studies in the Social Sciences, Stanford University, June 1975.
12. Uhler, R. S., "Costs and Supply in Petroleum Exploration: The Case of Alberta," *Canadian Journal of Economics*, February 1976.

UNCERTAINTY AND THE OPTIMAL SUPPLY POLICY FOR AN EXHAUSTIBLE RESOURCE

Geoffrey Heal, UNIVERSITY OF SUSSEX, ENGLAND

The rate at which a society should choose to deplete its stock of an exhaustible resource will naturally depend on the size of these stocks, the alternatives to the resource concerned in its various applications, and its value in use. Even if one had complete information about every one of these factors, the choice of a depletion policy would still be a matter of considerable technical complexity and political contentiousness, involving as it does difficult economic and technical considerations and political judgments about the balance to be struck between present and future needs. In fact, of course, one does not have anything like complete information about any of these factors: estimates of the relevant magnitudes are inherently uncertain, and this

Advances in the Economics of Energy and Resources, Vol. 2, pp. 119–147.
Copyright © 1979 by JAI Press, Inc.
All rights of reproduction in any form reserved.
ISBN: 0-89232-079-6

compounds the problems innate in an already difficult analysis. Our aim in this paper is to isolate particular classes of uncertainty, to propose persuasive yet tractable formalizations of these, to illustrate how such formalizations can be incorporated into models of the choice of a depletion rate, and to examine the consequences of so doing.

The uncertainties with which we shall be concerned, are those associated with the supply of the resource. We distinguish three "basic cases," which seem in combination sufficiently rich to encompass most of the relevant possibilities. These are:

Case 1. There are proven reserves of a resource whose magnitudes are known with certainty. There is in addition a possibility that further reserves will be discovered in the future, but the timing and magnitudes of such discoveries are uncertain.

Case 2. Again, there are proven reserves of a resource whose magnitudes are known with certainty. There is in addition a possibility that a close substitute for this resource will be developed in the future, but the timing of the discovery, and the scale on which the substitute will be available, are uncertain.

Case 3. There is substantial uncertainty about the size of the reserves of the resource whose location is currently known.

It is obviously not difficult to find examples that fit these various cases. Oil certainly illustrates the points at issue in the first two. We have available proven reserves whose magnitude is known with tolerable precision. There is also a clear possibility of discovering new fields in the future, but the timing and magnitudes of such discoveries must be inherently uncertain. This illustrates the type of situation to be formalized in Case 1. As mentioned, oil can also illustrate the second case, for there is currently a great deal of research directed toward the development of substitutes for traditional oil sources. The developments of shale oil reserves and of tar sands clearly come into this category, as does the development of potentially more economical sources of nuclear power such as breeder and fusion reactors, or of sources of solar power. It is hardly necessary to point out the uncertainties surrounding the outcomes of such endeavors, and hence the need for an analysis of a case such as Case 2. Shale oil and tar sands can also serve to illustrate our third case, for there is of course considerable uncertainty about the magnitudes of the currently known reserves of these: uranium reserves could be taken to illustrate the same point.

In the sections that follow, we shall present some very simple models that illustrate these various cases, and then characterize optimal depletion policies for these, examining in particular the impact of uncertainty on these policies. In all cases technical details of the derivations of the optimal policies will be isolated in the Appendix at the end of the paper, with the main text concentrating on the presentation and interpretation of results.

1. CERTAINTY—THE BENCHMARK CASE

When talking about "the effects of uncertainty," what one has in mind is of course a comparison of the optimal decision with that which would have been made, had all parameter values been certain. To facilitate such comparisons, we begin by setting out the solutions to our model under conditions of certainty.

We consider first the very simplest of cases, known in the literature as the "cake-eating problem." A society has a fixed and unaugmentable stock S of an exhaustible resource: the benefits accruing from consuming this at a rate C_t at time t are given by a concave differentiable payoff function $U(C_t)$, and the society's aim is to choose a consumption profile C_t, $0 \leq t \leq \infty$, so as to maximize

$$\int_0^\infty U(C_t)e^{-\delta t}dt \tag{1}$$

subject to the obvious limitations that

$$C_t \geqq 0, \text{ all t, and } \int_0^\infty C_t dt \leqq S. \tag{2}$$

In equation (1), $\delta > 0$ is a discount rate applied to future benefits: the objective is thus to maximize the present discounted value of present and future benefits, subject to obvious availability and non-negativity constraints. It is clear that the discount rate δ, which determines the relative weights of present and future benefits, will be a crucial parameter in the solution. One would of course expect an increase in δ, and thus a decrease in the relative importance of future benefits, to bias the consumption profile toward the present. The curvature of the payoff function $U(C_t)$ should also be important, as this determines the rate at which the marginal benefits from an extra unit of consumption decline: if this rate is high and the payoff function flattens rapidly with increasing C_t, one would expect this to discourage the concentration of consumption early in the plan period. To illustrate these points more formally, we consider a payoff function of the form

$$U(C_t) = \frac{C_t^{\eta+1}}{\eta+1} \tag{3}$$

where $\eta < -1$. This function has the property that the elasticity of the marginal payoff with respect to consumption is constant and equal to η: this elasticity may be written

$$\eta = \frac{\partial^2 U(C_t)}{\partial C_t^2} \frac{C_t}{\partial U(C_t)/\partial C_t}$$

and is of course just the Gaussian curvative of the function.

It is now routine to establish that the maximum of equation (1) subject to (2) must satisfy

$$\frac{\dot{C}_t}{C_t} = \frac{\delta}{\eta} \text{ or } C_t = \frac{-\delta}{\eta} S e^{\delta t/\eta} \tag{4}$$

where $\dot{C}_t = dC_t/dt$.

An intuitive justification for this result is readily available. A necessary condition for a path C_t to solve the problem is that there exists a constant P_0 such that

$$U'(C_t)e^{-\delta t} = P_0 \text{ for all } t \tag{5}$$

where $U'(C_t) = dU(C_t)/dC_t$. This simply requires the present value of the marginal payoff to be the same in all periods, and it is obvious that if this condition were not satisfied then some transfer of consumption from periods of low $U'(C_t)e^{-\delta t}$ to those of high values of this variable, would increase the value of the objective function. Differentiating equation (5) and rearranging leads to equation (4). Consumption thus declines exponentially at a rate controlled by the two parameters that one would expect to be important. An increase in δ leads to a consumption profile with a higher initial value and a more rapid rate of decline, while an increase in the absolute value of η, the rate at which marginal payoff declines, has an offsetting effect. As an illustration of the orders of magnitude that emerge from equation (4), note that a reasonable value for δ would be 10 percent p.a., or 0.10: values of η are more arguable, but most people would agree to be interested in a range of values that include -2.

In this case,

—consumption falls at 5 percent p.a.
—consumption at each date is at a rate of 5 percent of the existing stock
—the half-life of the stock is approximately fourteen years.

These results are clearly only illustrative, and indeed, by noting that doubling δ/η doubles the consumption rates and halves the half-life, the reader can easily conduct his own experiments. They do, however, make the point that a discount rate in the region of 10 percent p.a.—the figure often used by public authorities—will typically lead to a rapid decline of consumption levels and remaining stocks. This may of course tell us something about the appropriateness of such a discount rate.

We now turn briefly to some situations which are slightly more complex but nevertheless still certain. Society still has a given initial stock S_0: however, at a date T known with certainty a new stock of size S_1 (also known with certainty) will become available. The motivation for looking at such a situation is simple: this is the certain version of Case 1 presented above, and we shall subsequently permit both T and S_1 to be random variables.

But that analysis will be clearer if we already know the results of the present case. In fact, these results are very simple: at all dates except possibly T, the depletion policy must satisfy the equation

$$\frac{\dot{C}}{C} = \frac{\delta}{\eta} \tag{6}$$

and the resulting consumption profiles may look either as in Figure 1 or as in Figure 2.

The basic point to grasp here is that ideally one would like to have the present value of the marginal payoff the same at all times, and thus have an exponentially declining consumption level, as in Figure 1. However, complications arise because the operative constraints are now:

$$\int_0^T C_t dt \leqq S_0 \tag{7}$$

$$\int_T^\infty C_t dt \leqq S_1 + S_0 - \int_0^T C_t dt. \tag{8}$$

There are of course many paths that satisfy equations (6) and (7) from 0 to T: the highest of these will just exhaust S_0 at T, and will be associated with a consumption level we shall call C_T^- at T. Now, if this path is followed up to T, the relevant constraint from T will be

$$\int_T^\infty C_t dt \leqq S_1$$

and we will denote the initial consumption level at T on the highest path to satisfy this and equation (6) by C_T^+. Now, it is quite possible that for small S_0 or large T or S_1, $C_T^- < C_T^+$. In this case the optimal consumption path will have a jump in consumption from C_T^- to C_T^+ at T, as in Figure 2. Of course, in such a situation the marginal return to extra consumption just before T may be greatly in excess of that just after, but the constraint equation (7), reflecting the fact that the new stock is not yet available, prevents a transfer of consumption from after to before T to take advantage of this difference: there is an asymmetry in that consumption can be postponed from before to after T, but not brought forward from after to before. It is worth noting, before moving on, that Figure 1 shows a situation where there would be no social gain to advancing the date T at which the new reserves are available: in economic terms, they are already available as soon as needed. In Figure 2, however, the situation is different: there would be a clear gain to advancing T, and this would be proportional to the difference between the marginal payoffs before and after T.

As a final item in the certainty section, we consider a case which is the certain form of Case 2 above. Again there is a given initial stock S_0, and at a date T known with certainty a flow of a perfect substitute for the resource will become available at rate F. The real-world scenario to which

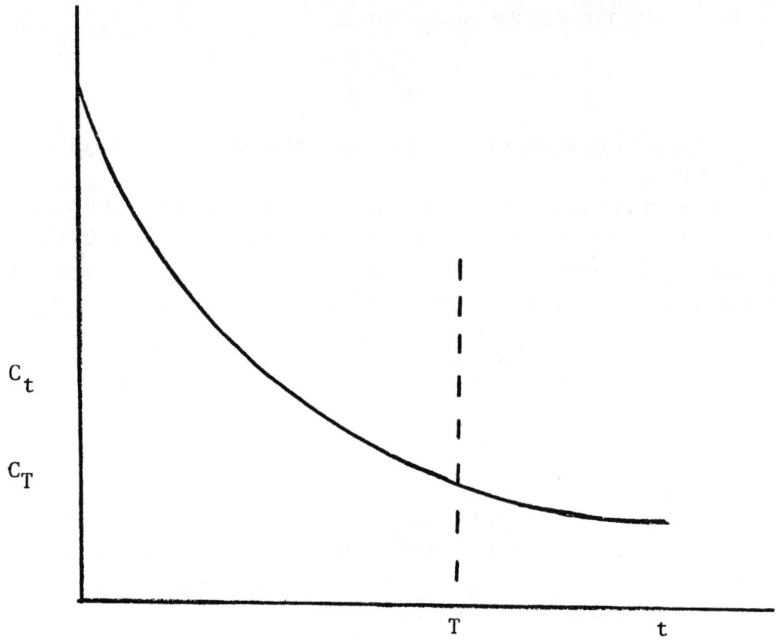

Figure 1.

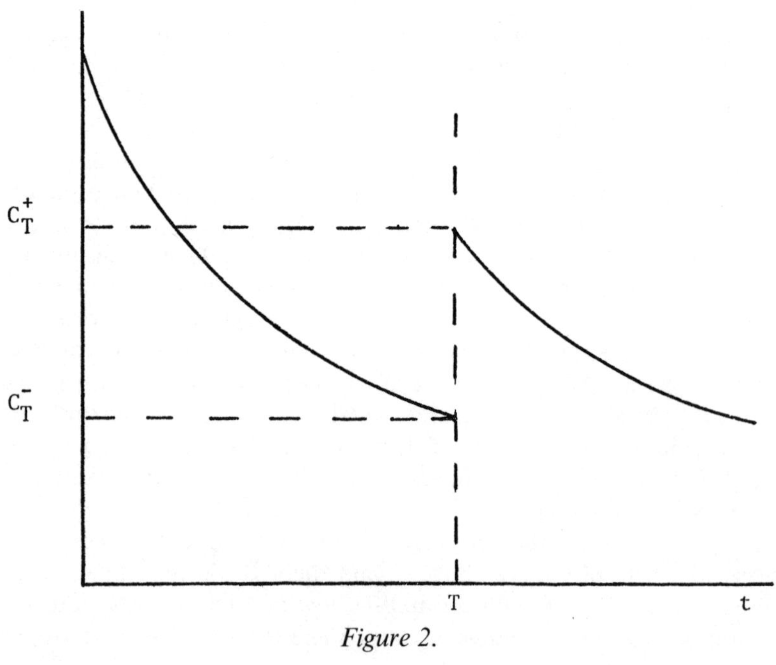

Figure 2.

this corresponds is, of course, that at T there is a technological innovation which leads to the development of an alternative source of whatever service the resource provides, at a rate bounded above by F. One could, for example, take S_0 to be fossil fuel, T the date at which solar energy will be available on a large scale and F to be the capacity of solar generators. In due course, we shall permit F and T to be random variables.

The solution to the problem as currently posed is quite straightforward: there exists a date $T_1, T_1 \geqq T$, such that:

$$\begin{aligned} &\text{for } 0 \leqq t \leqq T_1, \frac{\dot{C}_t}{C_t} = \frac{\delta}{\eta} \\ &\text{for } t \geqq T_1, C_t = F. \end{aligned} \tag{9}$$

The various possibilities permitted by equation (9) are shown in Figures 3 and 4. The operative

constraints are now

$$\begin{aligned} &\int_0^T C_t dt \leqq S_0 \\ &\dot{S}_t = F - C_t, t \geqq T, \text{with} \\ &S_T = S_0 - \int_0^T C_t dt \text{ and } S_t \geqq 0. \end{aligned} \tag{10}$$

Up to T, everything is as in the simpler cases: at T the system inherits whatever remains of the initial stock and has this augmented at a rate F. Consumption then has to be chosen to maintain a non-negative stock. To develop some feeling for the solution, suppose the stock inherited at T to be zero. Then the only options would be to consume the flow F as it arrives, or to store some of it for future consumption. But the discount factor insures that present consumption is valued at least as highly as future consumption: there is no merit in storing, and hence $C_t = F$ from T onward. Figure 4 illustrates a case where a path on which $C/C = \delta/\eta$ exhausts S_0 at T at a consumption level C_T less than F. Hence the stock left at T is zero and $C_t = F$ from then on. Of course, the economy would ideally like to bring consumption forward from after to before T, as it wished to in Figure 2: however, the constraints will not permit this and the optimum is associated with differences between the marginal payoffs just before and just after T. Figure 3 illustrates the case where S_0 is large enough, or T small enough, that there are paths satisfying $\dot{C}/C = \delta/\eta$ from 0 to T which arrive at T with $C_T > F$ and positive stocks: in such cases there is no discontinuity in consumption.

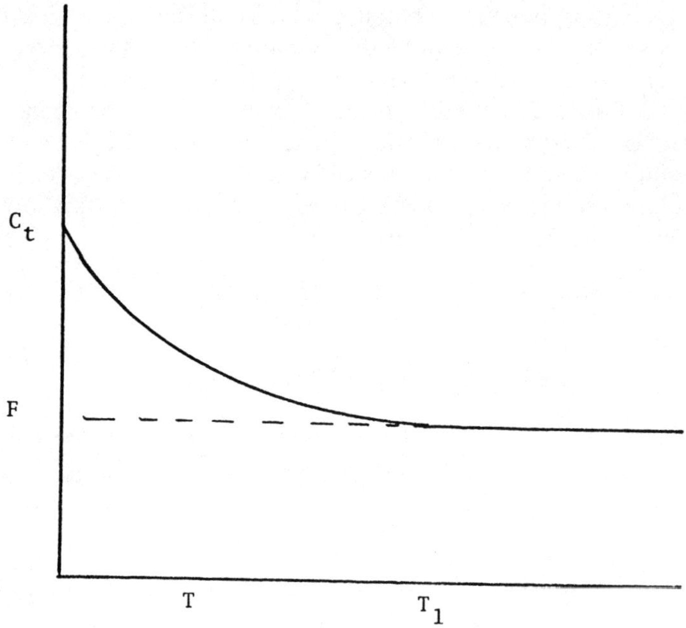

Figure 3.

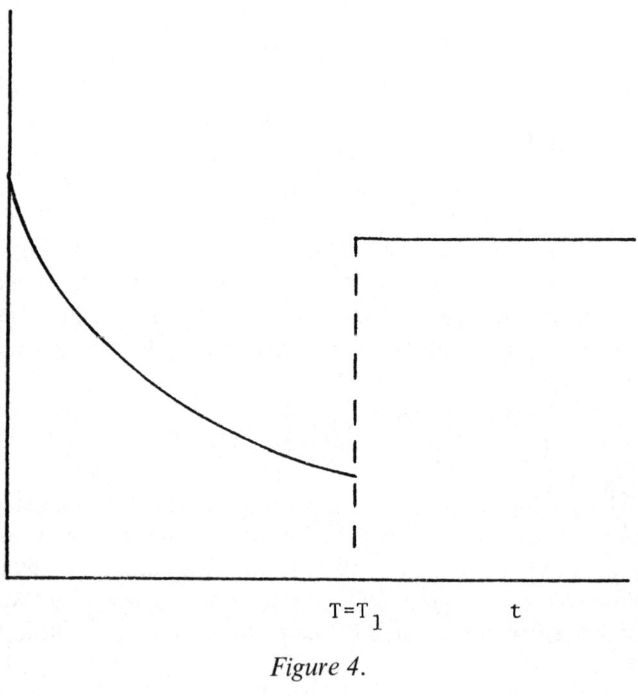

Figure 4.

Case I

We now turn our attention explicitly to uncertainty, and consider situations where a stock S_1 will be discovered at date T, with either S_1 or T unknown. Let S_T be given by

$$S_T = S_0 - \int_0^T C_t dt.$$

Then an important construct is the state valuation function $W(S_1 + S_T)$ given by

$$W(S_1 + S_T) = \underset{C_t, t \geqq T,}{\text{Maximum}} \int_T^\infty U(C_t)e^{-\delta(t-T)}dt \tag{11}$$

where the maximization is subject to

$$\int_T^\infty C_t dt \leqq S_1 + S_T.$$

$W(S_1 + S_T)$ thus gives the payoff, discounted back to time T, of discovering a stock of size S_1 when S_T remains of the initial stock. By standard theorems of dynamic programming, W( ) inherits many of the properties of U( ) – in particular its concavity.

T random, S_1 known

To keep the analysis sufficiently simple to be intuitively tractable, we shall consider in turn uncertainty about each single element of the problem. Initially we consider a situation where S_1, the size of the new discovery, is known, but T, the discovery date, is uncertain. This corresponds to a situation where one has *a priori* grounds of perhaps a geological nature for believing that further reserves of a resource must exist, but these have not been located. We suppose that there is probabilistic information about the discovery date, with a marginal density function w(T). The objective is now to maximize the expected payoff where the expectation is over dates T: this can be written as

$$\int_0^\infty w(T)\{\int_0^T U(C_t)e^{-\delta t}dt + W(S_1 + S_T)e^{-\delta T}\}dT \tag{12}$$

or, after integration by parts and setting

$$\int_t^\infty w(t)dt = \Omega(t),$$

$$\int_0^\infty \{U(C_t)\Omega(t)e^{-\delta t} + W(S_1 + S_t)w(t)e^{-\delta t}\}dt \tag{13}$$

and this has to be maximised subject to the restrictions

$$\int_0^\infty C_t dt \leqq S_0, C_t \geqq 0.$$

The solution to this problem consists of two parts: a strategy to be followed once T has occurred and S_1 is available (this is the strategy that achieves the maximum in equation (11)) and a conditional strategy to be pursued until T. Obviously this latter strategy is the one that is relevant under the conditions of uncertainty, and it is that that is analyzed below: the former is identical to the first certain case considered in the previous section. Defining

$$p(t) = w(t)/\Omega(t)$$

as the value of the density function at t is conditional upon the discovery not having been made by t, it can be shown that from 0 to T the optimal depletion policy must satisfy

$$\frac{\dot{C}_t}{C_t} = \frac{\delta}{\eta} + \frac{p(t)}{\eta}\left(1 - \frac{W'(S_1 + S_t)}{U'(C_t)}\right). \tag{14}$$

Equation (14) is clearly a generalization of the differential equation (4) derived for the certain case, and indeed if w(t) were to equal zero over an interval, the two equations would be identical on that interval. Typically, however, $p(t) > 0$, and also $U' > W'$: letting

$$\gamma_t = 1 - W'/U'$$

we have

$$\frac{\dot{C}_t}{C_t} = \frac{\delta + \gamma_t P(t)}{\eta}$$

and the effect of uncertainty is to augment the effective discount rate by an amount γP. For a simple distribution such as the Poisson,

$$w(t) = \alpha e^{-\alpha t},$$

one has $p(t) = \alpha$.

If in addition S_1 is large relative to S_0, W' is likely to be small relative to U' and we can approximate the above equation by

$$\frac{\dot{C}_t}{C_t} = \frac{\delta + \alpha}{\eta} \tag{15}$$

with solution

$$C_t = -\frac{(\delta + \alpha)}{\eta} S_0 e^{\frac{\delta + \alpha}{\eta}.t}$$

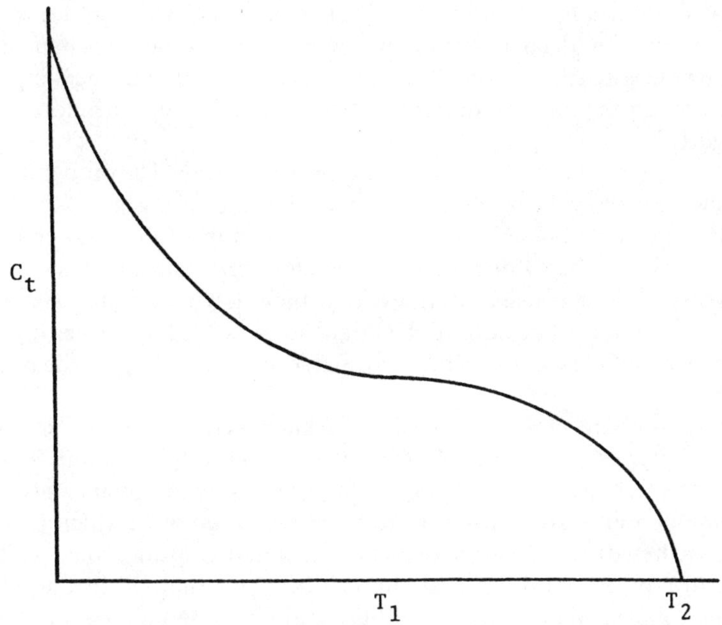

Figure 5.

This is again identical to equation (4), except for the augmentation of the discount rate. We thus see that the uncertain problem described by equations (12) or (13) admits under certain circumstances a solution identical to that of a certain problem derived from the original by assigning a zero probability to the uncertain event, and raising the discount rate. This is a rather powerful certainty-equivalence result, and we shall encounter variants of it at several points in our analysis. There is a rule-of-thumb in cost-benefit analysis to the effect that it may be appropriate to allow for uncertainty by discounting the future more heavily: this result illustrates a set of conditions under which that rule is acceptable.

Let us now return to equation (14) and examine the nature of its solution in some other special cases. Continue to suppose that S_1 is large relative to S_t, so that W'/U' can be neglected: then

$$\frac{\dot{C}_t}{C_t} = \frac{\delta + P(t)}{\eta} \text{ where } P(t) = w(t)/\Omega(t).$$

Now suppose that P(t) is positive only on an interval $T_1 \leqq t \leqq T_2$: there are definite upper and lower bounds on the dates at which the discovery might be made. Now

$$\Omega(t) = \int_t^\infty w(t)dt = \int_t^{T_2} w(t)dt$$

so that P(t) tends to infinity as $t \rightarrow T_2$. η is of course negative: we thus see that consumption decreases at an ever greater rate as the right-hand end of the interval is reached. For $t \leqq T_1$, however, $P(t) = 0$ and one simply has the customary exponential decline. A depletion path might therefore look as in Figure 5.

There is a straightforward intuitive explanation of this. The probability that the new reserves will have been discovered is increasing in t between T_1 and T_2, and is of course 1 by T_2. The chances of the right-hand end of the above curve ever being operative are therefore very small, and the optimal policy assigns low consumption levels to these low-probability states as a minimal insurance. The bulk of the initial stock is used up earlier on, when the chances of the new stocks being available are low and hence the resource is still scarce.

Before considering the reverse case of T known and S_1 random, it is worth considering briefly the effect of a change in the uncertainty about T. It can be shown (perhaps not surprisingly) that a mean-preserving increase in the uncertainty about T(i.e., a spread of the distribution w(T) which leaves its mean unaltered) will cause a reduction in initial consumption—and vice versa. At a purely intuitive level, this can best be visualized by considering Figure 5, and imagining the initial distribution to be uniform on T_1, T_2. Then imagine a new uniform distribution centered on T_1, T_2 but positive on a larger interval: it is not difficult to see that a path with the same general characteristics as that in Figure 5 (initially exponential, zero at the right-hand end of the interval) must start with a lower initial consumption level.

T known, S_1 random

We have so far considered a situation where a known stock will be discovered at an unknown date. We now consider the converse—a stock of unknown size whose discovery date is certain. This could perhaps correspond to a situation where intensive prospecting is due to be carried out in a particular region, during a particular period: this is virtually certain to lead to some discovery, but the exact size cannot be predicted. Probabilistic information about possible discovery sizes is available, and is summarized in a marginal density function $w(S_1)$. $W(S_1 + S_T)$ is defined as in equation (11) before, so that the problem is to maximize

$$\int_0^T U(c_t)e^{-\delta t}dt + \int w(S_1)W(S_1 + S_T)e^{-\delta T}dS_1 \tag{16}$$

subject to $\int_0^T C_t dt \leqq S_0, C_t \geqq 0$.

The second term in the maximand is of course the expectation of the state valuation function W( ) over all possible values of S_1: from now on the expectation operator will just be denoted by E. The conditions that a solution must satisfy are fairly simple in this case: they are

$$\frac{\dot{C}_t}{C_t} = \frac{\delta}{\eta} \text{ for } 0 \leqq t \leqq T, \tag{17}$$

$$U'(C_T) \geqq EW'(S_1 + S_T). \tag{18}$$

Equation (17) is familiar by now and needs little comment: consumption declines at the usual exponential rate from 0 to T. Equation (18) is also relatively obvious in its implications. If it were possible to shift consumption both forward and backward in time, it would be satisfied with equality. However, because of the asymmetry set up by the inability to do this (and already noted in the certain case), there may be an inequality here, with the marginal payoff to consumption just before T exceeding the expected value of that just after. One can in fact be a little more precise about the circumstances when strict inequality may hold. Suppose that $S_T = 0$, and consider the solution to the problem

$$\text{Maximize} \int_T^\infty U(C_t) e^{-\delta(t-T)} dt$$

$$\text{subject to} \int_T^\infty C_t dt \leqq S_1 .$$

If $S_T = 0$, the solution to this problem describes the consumption path from T on. Call the initial level of consumption on such a path C_T^+: then corresponding to each S_1 is a value C_T^+. Now denote by C_T^- the value of C_T on that path from 0 to T which satisfies equation (17) and just exhausts S_0 (i.e. leaves $S_T = 0$). Suppose finally that it is known with certainty that the stock discovered at T will be at least $\underline{S}_1$, and the $\underline{C}_T^+$ denote the corresponding C_T^+. Then it should be clear that

$$C_T^- < \underline{C}_T^+ \text{ implies } U'(C_T) > EW'(S_1 + S_T).$$

Figure 6 illustrates this point. AB is a path on which $\dot{C}/C = \delta/\eta$, and which exhausts S_0 at T. The range of possible initial consumption levels from T on is given by CD. This is obviously the analogue of Figure 2 in the certain case, and there is a corresponding difference in *ex ante* marginal valuations before and after T. Before moving on to consider other cases, it is worth noting that a change in the distribution of S_1 that keeps the solution in this regime will have no effect on consumption from 0 to T. Thus, increases in D, decreases in C toward but not beyond B, or changes in the probabilities of points within the range CD will all leave AB completely unaltered. The

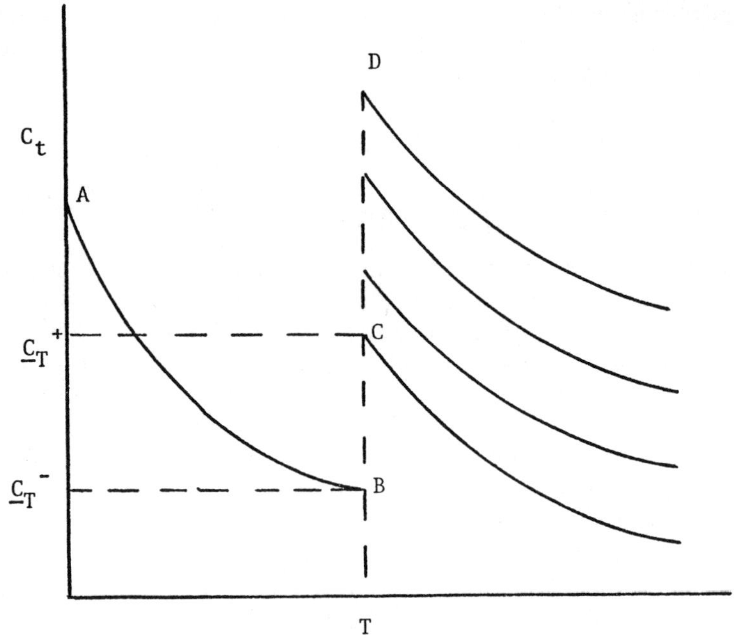

Figure 6.

binding nature of the constraint at T insulates the initial stage of the problem from the final one, so that changes in the uncertainty regarding the latter have no effect on the former.

Of course, if equation (18) holds with equality and $U'(C_T) = EW'(S_1 + S_T)$, then matters are a little different. And this would certainly be the case in a situation where there was a positive probability that $S_1 = 0$ and T brings no further discovery. In such a situation *ex ante*, or planned, marginal payoffs are equated, at least in expected value terms, before and after T, though *ex post* they will typically diverge. This equality links events between 0 and T to those after T in a way which can readily be analyzed. For example, suppose that there is a mean-preserving increase in the uncertainty about possible values of S_1 – that is, the distribution $w(S_1)$ changes in a way which preserves the mean but increases measures of its spread. Then if the derivative W′ of W is a convex function (the most likely case, as it is bounded below by zero), it is easy to show that such an increase in uncertainty about S_1 leads to a decrease in the consumption levels chosen at all dates from 0 to T. Conversely, a decrease in the uncertainty associated with the estimate of S_1 will lead to an increase in consumption from 0 to T. This is obviously very much in accord with intuition.

Case 2

We now turn our attention to the second of the sources of uncertainty mentioned in the introduction. We are interested in a situation where there is a known initial stock of a resource, and there is in addition a possibility that a close substitute will be developed in the future. There is however, uncertainty about the timing of this discovery and the scale on which the substitute will be available. As in our discussion of the certain version of this problem, we let T be the date of the discovery and F be the rate at which the substitute is available from T on. Of course, once T has occurred, both F and the stock available, S_T, are certain: the problem then corresponds to one of the certainty problems already solved. We therefore focus our attention on the choice of a policy from 0 to T.

F certain, T random

As before, we shall consider the effect of uncertainty about one element of the problem at a time. Initially we concentrate on the impact of uncertainty about the timing of the innovation, assuming that a reasonably precise estimate of its output is already available. We shall assume for simplicity that any of the initial stock remaining at time T contributes an insignificant amount to consumption levels from T on: post-T, consumption possibilities are dominated by F. Letting w(T) be the marginal density function describing the discovery probabilities, the objective is clearly to maximize.

$$\int_0^\infty w(T)\left\{\int_0^T U(C_t)e^{-\delta t}dt + \frac{U(F)e^{-\delta t}}{\delta}\right\}dT. \tag{19}$$

Equation (19) has an obvious similarity to (12), except that W takes, by assumption, a rather simple form in this case. Maximizing equation (19) subject to the integral constraint on resource consumption between 0 and T leads to the following differential equation for C_t:

$$\frac{\dot{C}_t}{C_t} = \frac{\delta + P(t)}{\eta} \tag{20}$$

where $p(t) = w(t)/\Omega(t), \Omega(t) = \int_t^\infty w(t)dt$.

Equation (20) is of course identical in general form to (14) and the equations derived from it: indeed, the assumption that consumption from the initial stock is insignificant after T would imply that $W' = 0$ in equation (14), in which case it is identical with (20). Conversely, allowing consumption from the initial stock to matter after T would in the present case lead to equation (14) rather than (20). The justification for going straight to the simple case here is thus, in part, that the complex one has already been viewed, and also that the unimportance of remnants of the initial stock seems a more reasonable assumption when T brings an entirely new technology than when it merely brings more of the initial stock.

Because of the obvious similarity between the effects of uncertainty about T in Cases 1 and 2, it is not really necessary to pursue the analysis much further. Equation (20) makes it clear that we once again have a certainty equivalence result: the solution to our problem could be obtained by assigning a zero probability to the uncertain event and solving the resulting certain problem, but with a higher discount rate. As before, if the distribution w(t) is Poisson, with

$$w(t) = \alpha e^{-\alpha t},$$

then $p(t) = \alpha$ and the equivalent certain problem has a constant discount rate. The other possibility considered in Case 1 was that w(t) was zero except on an interval, (T_1, T_2): the effects of such an assumption would also be the same here as there, with consumption levels falling sharply as T_2 is approached, but falling exponentially up to T_1—as shown in Figure 5. As in the equivalent version of Case 1, it can be shown that a mean-preserving increase in uncertainty about T will lead to a reduction in the initial consumption level.

F random, T certain

We now turn to the converse situation, where the date of the new technology's availability is certain but its scale is not. This will again turn out to be very similar to the corresponding version of Case 1. As in that case, the depletion policy satisfies

$$\frac{\dot{C}_t}{C_t} = \frac{\delta}{\eta}$$

from 0 to T: in addition, it is required that

$$U'(C_T) \geqq EU'(F) \tag{21}$$

where the expectation is over possible values of F. Equation (21) is clearly analogous to (18), and holds with equality or inequality in the same sorts of situations. Thus, suppose that there is a lower bound F on possible values of F, and that C_T^- is again the consumption level at T on a path from 0 to T that exhausts S_0 at T. Then if

$$C_T^- < F,$$

one has strict inequality in equation (21): the difference in marginal payoffs at two adjacent dates is sustained because of the binding constraint. Figure 7 illustrates such a case: AB is a path satisfying $C/C = \delta/\eta$ and exhausting S_0

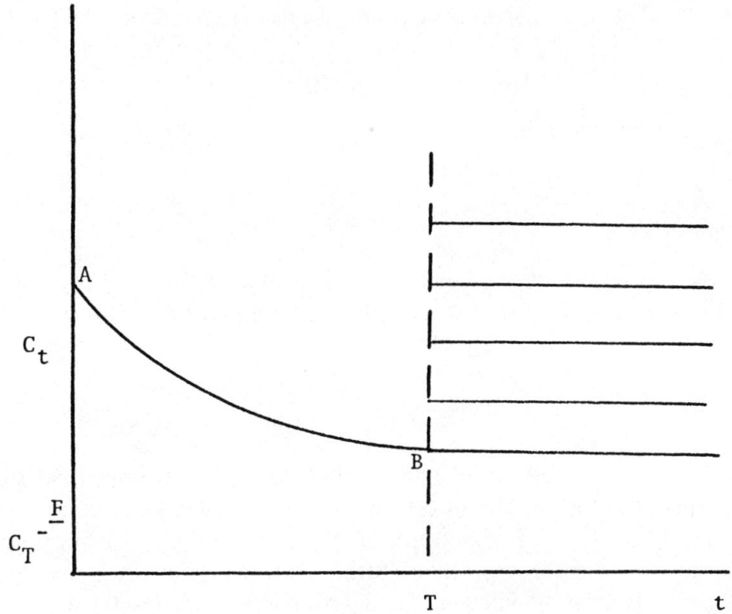

Figure 7.

at T, and the range of possible F values starts at F. It is clear that, as in the corresponding version of Case 1, a change in the distribution of F values which maintains a regime such as that shown $(F > C_T^-)$ will have no effect on the choice of a consumption path from 0 to T: the binding constraint again disconnects the two phases of the solution.

The alternative possibility is that equation (21) should hold with equality: a sufficient condition for this is clearly that there is a positive probability that $F = 0$. There is, in other words, a risk that the new technology will be a complete failure. Such a situation is obviously incompatible with our earlier simplifying assumption that the remains of the initial stock make an insignificant contribution to consumption after T, so that it is necessary to revise the analysis slightly be defining a function

$$W(F,S_T) = \text{Maximum} \int_T^{\infty} U(C_t)e^{-\delta(t-T)}dt$$

subject to S_T given, $S_t \geqq 0$,

$$\dot{S}_t = F - C_t, t \geqq T.$$

W is again a state valuation function, corresponding to the solution to the certain problem described in equation (10). It gives the maximum discounted payoff obtainable from T on, with a flow at rate F and a stock at T of S_T.

The overall problem can now be posed as one of maximizing

$$\int_0^T U(C_t)e^{-\delta t}dt + EW(F,S_T)e^{-\delta T}$$

subject to

$$S_0 \geqq \int_0^T C_t dt \; and \; S_T, C_t \geqq 0$$

where the expectation is of course over values of F. In this case, the solution, in addition to satisfying the usual differential equation $\dot{C}/C = \delta/\eta$, has to satisfy

$$U'(C_T) = E\partial \frac{W(F,S_T)}{\partial S_T} \tag{22}$$

Equation (22) links the initial and final stages of the problem, so that a change in the nature of the uncertainty about F affects the consumption path chosen from 0 to T. For example, if $\partial W/\partial S$ is a declining convex function of F (the most reasonable case) then a mean-preserving increase in uncertainty about the values of F will lead to a reduction in C_t for $0 \leqq t \leqq T$, and conversely a decrease in uncertainty, with mean constant, will lead to an increase. This is exactly as one would expect, and in accordance with the equivalent results in Case 1.

Case 3

We turn now to the type of uncertainty exemplified by Case 3—undoubtedly the simplest of all to describe, but the most complex to analyze. The problem, then, is to describe the optimal depletion policy when the size of the currently available stock is unknown. One can think of the pure case of this sort as extracting oil from reserves whose geological structure is so complex that estimates of the total amount recoverable are subject to large errors until the final barrel has been extracted. Perhaps an even purer form of the problem would arise from draining gasoline from an opaque tank whose contents cannot be viewed and whose initial level is unknown. It is clear that the date at which a given depletion policy exhausts the stock depends on that stock, and hence in the present case is also uncertain: corresponding to any given depletion policy there is therefore a distribution of possible exhaustion dates. Selecting a depletion policy implies not only selecting a consumption profile, but also a distribution of dates at which consumption will cease.

In order facilitate an understanding of this problem, we shall begin by reviewing the solution to the "nearest" certain problem—a problem slightly different in structure from any of the certain cases analyzed before. The

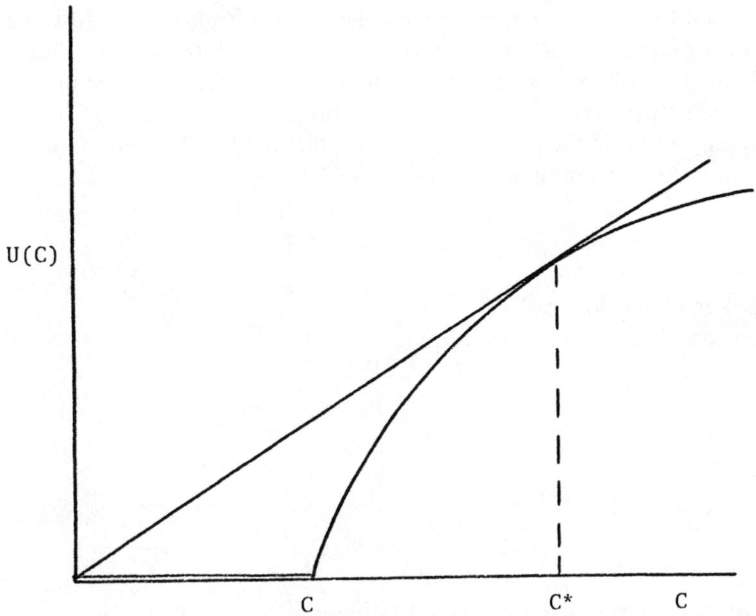

Figure 8.

crucial difference is in the form of the payoff function, which we shall suppose for the moment to have the shape shown in Figure 8. There is a consumption level $\underline{C}$ such that $U(C)=0$ for $C \leqq \underline{C}$: for $C \geqq \underline{C}$, U( ) is a well-behaved concave function. The implication is that there is some minimum level of input needed before any payoff can be achieved—there is an element of indivisibility in consumption. In many applications this would clearly be a very reasonable assumption.

Now consider the solution to the problem:

$$\text{Maximize } \int_0^\infty U(C_t)e^{-\delta t}dt$$

$$\text{subject to } \int_0^\infty C_t dt \leqq S_0$$

when the function U( ) is as shown in Figure 8. The solution we derived earlier for a payoff function of the form

$$U(C) = \frac{C^{\eta+1}}{\eta+1}$$

involved consumption falling exponentially to zero. Clearly that cannot be optimal in the present case, for associated with any such path is a date *T* such that for $t > T\, C_t < \underline{C}$, and all consumption from T onward is wasted: it contributes nothing to the payoff. One would thus surmise that in the present case the optimal policy will involve consumption declining until it reaches a strictly positive critical value (presumably not below $\underline{C}$) and then dropping to zero. Such a surmise would be correct: the optimum involves consumption declining according to

$$\frac{\dot{C}_t}{C_t} = \frac{\delta}{\eta}, 0 \leqq t \leqq T$$

and then $C_t = 0$ for $t > T$.
T is a date determined by the requirements that

$$\int_0^T C_t dt = S_0$$

and

$$\frac{U(C_T)}{C_T} = U'(C_T). \tag{23}$$

All of this is conventional except for equation (23): this defines the level of consumption at which the stock is to be exhausted and consumption is to cease. This level is characterized by equality of the average and marginal payoff rates. The average rate is of course given by the slope of a line from the origin to U(C), C and the marginal by the slope of U( ): the relevant consumption level is thus that denoted by C* in Figure 8, and the optimal policy involves an exponential decline of consumption to C*, with the stock just exhausted then and zero consumption thereafter.

Condition (23), though puzzling at first sight, can in fact be given a very simple interpretation. Let T be the date at which $C = C^*$ on an optimal policy. Suppose that consumption just prior to T is reduced by one unit: then the loss of payoff is $e^{-\delta T} U'(C_T)$. The unit of consumption thus saved can be used to prolong slightly the period during which consumption is positive. If consumption were to be held constant at its terminal rate during this very short prolongation (and it would certainly be possible to improve upon any alternative policy), the extra time bought would be $1/C_T$, and the utility gain would be $e^{-\delta T} U(C_T)/C_T$. Obviously a necessary condition for T and C* to be chosen optimally is that the net gain from such an extension should be zero: hence (23).

Let us now return to the uncertain case. The total initial stock S_0 is uncertain, and this uncertainty is described by a marginal density function $w(S_0)$. On any given consumption path C_t, consumption to date t′ is $\int_0^t C_t\, dt$ and hence the marginal density function describing the likelihood of exhaustion at t′ is w $(\int_0^t C_t dt)$. We shall let

$$Z_{t'} = \int_0^{t'} C_t dt$$

so that we can write $w(Z_{t'})$ as the probability of exhaustion at t′ on path C_t. The overall problem is thus to maximize

$$\int_0^{\infty} w(Z_{t'}) \left\{ \int_0^{t'} U(C_t) e^{-\delta t} dt \right\} dZ \tag{24}$$

subject to $\frac{dZ}{dt} = C_t$.

Application of standard techniques leads to the conclusion that a solution to this problem must satisfy the equation

$$\frac{\dot{C}_t}{C_t} = \frac{\delta}{\eta} + \frac{p(Z_t)}{\eta} \frac{(U'(C_t) - U(C_t)/C_t)}{U'(C_t)} \tag{25}$$

where $p(Z_t) = w(Z_t) / \int_{z_t}^{\infty} w(Z) dZ$.

Obviously there is much that is familiar in this equation. For $p(Z_t) = 0$ (a zero conditional probability of exhaustion at the current cumulative consumption), it is just the standard exponential decay at rate δ/η. And of course the term in parentheses in equation (25) is just the difference between the right- and left-hand sides of (23). If the total initial stock were exactly Z_t, this expression would therefore give the negative of the return to postponing a marginal increment of consumption and using it to extend the period of positive consumption. Multiplied by $p(Z_t)$, it gives the expected return to such postponement. Rearranging (25), it is clear that the level of consumption

$$\frac{\dot{C}_t}{C_t} = \frac{p(Z_t)}{|\eta|} \frac{(U(C_t)/C_t - U'(C_t))}{U'(C_t)} - \frac{\delta}{|\eta|} \tag{25'}$$

on an optimal path changes at a rate proportional to the difference between the expected return to postponing consumption, and the cost of such postponement, which is measured by the discount rate δ.

In addition to giving an intuitive rationalization of equation (25), it is possible to derive certain general quantitative features of a solution. Thus initial consumption levels will typically be above C*, so that $U' < U(C)/C$: the optimal policy will therefore typically involve a consumption profile that declines more slowly than an exponential initially. In the long run, the converse may be true: C may decline below C* so that $\dot{C}/C$ becomes more negative than δ/η. By comparison with the other cases we have seen, a particularly distinctive feature of equation (25) is that it may produce a path on which consumption rises over some interval. In particular, if the average payoff level exceeds the marginal and the conditional probability of exhaustion is high, we may have $\dot{C}/C > 0$. For a payoff function such as that shown in Figure 8, the average could exceed the marginal if and only if $C > C^*$: for a more conventional function with $U(O) = 0$, the average would always exceed the marginal. (Note that a payoff function $C^{\eta+1}/\eta + 1$, for $\eta \leqq -1$, would not be sensible in the present context though it proved convenient in Cases 1 and 2. The point is that in this case $U(O) = -\infty$: but in the present problem C becomes zero with a positive probability on any path, so that all paths would have a payoff of $-\infty$.) The fact that imminent exhaustion should serve to raise the level of consumption, is certainly counterintuitive. However, it is worth recalling that in Case 1 (T random, S_1 known) imminent replenishment of the stocks had the reverse effect. Figure 5 shows a situation where new reserves will certainly be discovered by date T_2, and the optimal consumption profile conditional on nondiscovery actually decline toward T_2. In the present situation, of course, we have the reverse of this: imminent exhaustion (rather than replenishment) produces an increase (not decrease). At least these two counterintuitive results are consistent with each other.

There is at least one special case in which (25) can be integrated explicitly. Suppose that

$$U(C) = \frac{C^{\eta+1}}{\eta+1}, 0 > \eta > -1$$

(so that $W(0) = 0$) and that

$$w(Z) = \alpha e^{-\alpha Z}$$

so that $p(Z) = \alpha$. Then the integral of equation (25) is

$$C_t = \frac{(\delta/\eta)e^{\delta t/\eta}}{K - \dfrac{\alpha}{1+\eta}e^{\delta t/\eta}} \tag{26}$$

where $K < 0$ is a constant of integration. Clearly C behaves like an exponential with power δ/η in the long run, but initially it falls more slowly than an exponential.

Because of the complexity of Case 3, it is difficult to make clear statements about the effects of changes in uncertainty. There is however one result that is available—and again it is a somewhat surprising one. This is that a mean-preserving increase in the uncertainty about the initial stock leads to an *increase* in the initial level of consumption.

SUMMARY AND CONCLUSIONS

In the above analysis, we have considered the impact on the choice of a depletion policy of (a) uncertainty about the timing and magnitude of future resource discoveries; (b) uncertainty about the timing and magnitude of future technological developments; and (c) uncertainty about the size of the initial stock of the resource.

Although each case is individually quite simple, it seems fair to say that between them these cases cover most of the possibilities one is likely to encounter in practice, or at least serve as building blocks from which more complex cases can be constructed. Given the scope of the coverage, the results seem encouragingly consistent and simple. They are drawn together in a convenient summary form in the table.

Certain points emerge very clearly from this table. One is that the impact of uncertainty about a significant future date is the same in both of Cases 1 and 2: the effect discount rate is raised, and a mean-preserving increase in uncertainty lowers C_0. Likewise, the effect of uncertainty about the size of a future discovery, be it of a stock S_1 or of a flow F, is similar in the two cases. The consumption path from 0 to T is identical in form to the deterministic case, and a change in the uncertainty surrounding S_1 or F may or may not influence this path depending upon whether or not the constraint relaxed at T is binding. A binding constraint effectively disconnects the two parts of the problem, so that a change in the nature of the uncertainty has no impact. The least predictable case is clearly the third one, which has little in common with the others.

TECHNICAL APPENDIX

The earlier discussions were largely heuristic and intuitive, and concerned with results rather than with derivations. In this Appendix we present concise proofs of the main propositions. Results not derived here can readily be obtained by extensions of the analysis or as special cases. Thus no results are given for the certain cases, as these are obviously all special cases of their stochastic equivalents.

Case 1

We consider first the case of random T and known S_1. Let

$$S_t = S_0 - \int_0^t C_\tau d^\tau,$$

and

$$W(S_1 + S_T) = \text{Maximum} \int_T^\infty U(C_t)e^{-\delta t}dt$$

subject to

$$\int_T^\infty C_t dt \leqq S_1 + S_T.$$

	Case 1		Case 2		Case 3
	T random S_1 known	T known S_1 random	T random F known	T known F random	Initial stock uncertain
Equation governing depletion policy	$\frac{\dot{C}}{C} = \frac{\delta}{\eta} + \frac{P}{\eta}\left(1 - \frac{U'}{W'}\right)$ Reduces to $\frac{\dot{C}}{C} = \frac{\delta + P}{\eta}$	$\frac{\dot{C}}{C} = \frac{\delta}{\eta}, O \leqq t \leqq T$ $U'(C_T) \geqq EW'$	$\frac{\dot{C}}{C} = \frac{\delta + P}{\eta}$	$\frac{\dot{C}}{C} = \frac{\delta}{\eta}, O \leqq t \leqq T$ $U'(C_T) \geqq EW'$	$\frac{\dot{C}}{C} = \frac{\delta}{\eta} + \frac{P}{\eta}\left(\frac{U' - U/C}{U'}\right)$
Effect of uncertainty	To increase effective discount rate. M.P.S. on distribution of T lowers C_O.	M.P.S. on distribution of S_1 may lower C_0 or have no effect, depending on whether constraint at T binds.	To increase effective discount rate. M.P.S. on distribution of T lowers C_0.	M.P.S. on distribution of F may lower C_0 or have no effect.	Nature of the consumption path is changed. M.P.S. on distribution of S_0 may raise C_0

NOTE: M.P.S. = mean-preserving spread or increase of uncertainty.

Then the problem is to

$$\text{Maximize} \int_0^\infty w(T)\left\{\int_0^T U(C_t)e^{-\delta t}dt + W(S_1 + S_T)e^{-\delta T}\right\}dT$$

$$\text{subject to} \int_0^\infty C_t dt \leqq S_0.$$

$$\text{Letting } \Omega(t) = \int_t^\infty w(\tau)d\tau,$$

the maximand can be written as

$$\int_0^\infty \{U(C_t)\Omega(t)e^{-\delta t} + w(t)W(S_1 + S_t)e^{-\delta t}\}\,dt.$$

We replace the integral constraint by the conditions

$$\dot{S}_t = -C_t, \quad S_t \geqq 0 \quad \text{all } t$$

and apply Pontryagin's Maximum Principle to the Hamiltonian

$$H = U(C_t)\Omega(t)e^{-\delta t} + w(t)W(S_1 + S_t)e^{-\delta t} - q(t)e^{-\delta t}C_t$$

where $q(t)$ is a co-state variable. Necessary conditions for optimality are

$$U'(C)\Omega(t) = q(t) \tag{A1}$$

and

$$\dot{q} - \delta q = -w(t)W'(S_1 + S_t). \tag{A2}$$

Differentiating and rearranging in (A1) and (A2), and recalling that

$$\frac{U''C}{U'} = \eta$$

yields

$$\frac{\dot{C}_t}{C_t} = \frac{\delta}{\eta} + \frac{P(t)}{\eta}\left(\frac{1 - W'(S_1 + S_t)}{U'(C_t)}\right)$$

which is equation (14) of the main text. Given this equation, the remaining statements in the text about the (T random, S_1 certain) case can readily be verified. Consequently we turn to the converse case.

The problem here is to

$$\text{Maximize} \qquad \int_0^T U(C_t)e^{-\delta t}dt + \int w(S_1)W(S_1 + S_T)e^{-\delta T}dS_1$$

$$\text{subject to} \qquad \int_0^T C_t dt \leqq S_0, \quad C_t \geqq 0.$$

Again we replace the integral constraint by the conditions

$$\dot{S}_t = -C_t, \quad S_T \geqq 0$$

and apply the Maximum Principle to the Hamiltonian

$$H = U(C_t)e^{-\delta t} \quad - q(t)e^{-\delta t}C_t$$

which in this case requires that

$$\begin{aligned} &U'(C) = q(t) \\ &\dot{q} - \delta q = 0 \\ &q(T) \geqq EW'. \end{aligned}$$

It is now a simple matter to make these equations yield (17) and (18) of the text. The statements in the text about the impact of a mean-preserving increase in uncertainty follow from the fact that if $W'(S_1)$ is a convex function and distribution f' is obtained from distribution f by a mean-preserving spread, then

$$\int W'(S_1)f(S_1)dS_1 \geqq \int W'(S_1)f'(S_1)dS_1,$$

with strict inequality if $W'(S_1)$ is strictly convex. [See for example Rothschild and Stiglitz (7).]

Case 2

Consider first the situation of T random, F known. Let

$$W(F,S_T) = \text{Maximum} \int_T^\infty U(C_t)e^{-\delta t}dt$$

subject to $\dot{S}_t = F - C_t, t \geqq T,$

and $S_t, C_t \geqq 0$

Then the overall problem is to

Maximize $$\int_0^\infty w(T)\left\{\int_0^T U(C_t)e^{-\delta t}dt + W(F,S_T)e^{-\delta T}\right\}dT$$

subject to $$\int_0^\infty C_t dt \leqq S_0.$$

The maximand can be integrated by parts to give

$$\int_0^\infty \{U(C_t)\Omega(t) + w(t)\,W(F,S_t)\}e^{-\delta t}dt$$

and the integral constraint replaced by the differential equation $\dot{S}_t = -C_t$

with the condition $S_t \geqq 0$. The relevant Hamiltonian is then

$$H = U(C_t)\Omega(t)e^{-\delta t} + w(t)W(F,S_t)e^{-\delta t} - q(t)e^{-\delta t}C_t$$

and necessary conditions for a solution are

$$U'(C)\Omega(t) = q(t)$$

$$\dot{q} - \delta q = -w(t)\frac{\partial W}{\partial S_t}.$$

Differentiating the first of these and substituting into the second yields

$$\frac{\dot{C}}{C} = \frac{\delta}{\eta} + \frac{P(t)}{\eta}\left(\frac{(1 - \partial W/\partial S_t)}{U'}\right)$$

from which the results in the text may be derived. The case of known T and random F is formally identical to the equivalent part of Case 1, and does not merit separate analysis.

Case 3

The basic problem is to

Maximize $$\int_0^\infty w(Z_T)\left\{\int_0^T U(C_t)e^{-\delta t}dt\right\}dZ \qquad \text{(A3)}$$

subject to $$\dot{Z} = C.$$

Problem (A3) can be rewritten as

$$\int_0^\infty w(Z_T)\frac{dZ_T}{dt}\left\{\int_0^T U(C_t)e^{-\delta t}dt\right\}dT$$

$$= \int_0^\infty w(Z_T)C_T\left\{\int_0^T U(C_t)e^{-\delta t}dt\right\}dT$$

$$= \int_0^\infty U(C_T)e^{-\delta t}\left\{\int_T^\infty w(Z_t)C_t dt\right\}dT.$$

Let $$\int_T^\infty w(Z_t)C_t dt \equiv \int_{Z_T}^\infty w(Z_t)dZ = \Omega(Z_t).$$

Then (A3) equals

$$= \int_0^\infty U(C_t) \quad \Omega(Z_t)e^{-\delta t}dt.$$

The relevant Hamiltonian is therefore

$$H = U(C)\,\Omega(Z)e^{-\delta t} + q(t)e^{-\delta t}C_t$$

and necessary conditions for a solution are

$$U'(C_t)\Omega(Z_t) = -q(t) \tag{A2}$$

and

$$\dot{q} - \delta q = U(C_t)W(Z_t). \tag{A5}$$

Differentiating (A4) and substituting into (A5) yields

$$-U''\dot{C}\Omega + U'p(Z)C_t = -\delta U'\Omega + U(C)p(Z)$$

which on rearrangement leads to equation (25) of the text and the associated results.

BIBLIOGRAPHICAL NOTE

The simplest version of the cake-eating problem under conditions of certainty has been discussed in several places: the original reference is Gale (2), and it is also discussed at some length in Heal (4). These authors are concerned primarily with its implications for the existence of an optimal policy—for the interesting feature of such a problem is that if $\delta = 0$ it can be shown to have no solution. One can see this immediately from equation (4) in the text, which presents the solution

$$C_t = -\frac{\delta}{\eta} S e^{\delta t - \eta}.$$

Clearly this makes little sense for $\delta = 0$. The implications of this problem for resource-depletion policies were examined in some detail by Dasgupta and Heal (1), who also present and analyze in some detail the models from which Case 1 and 2 have been developed. Case 3 seems first to have been analyzed by Kemp (1976), and then by Gilbert (1976) and by Loury (1976), although the present derivation is original and rather simpler than its predecessors.

REFERENCES

1. Dasgupta, P. S., and G. M. Heal, "The Optimal Depletion of Exhaustible Resources," *Review of Economic Studies*, Symposium, 1974.
2. Gale, D., "On Optimal Development in a Multi-Sector Economy," *Review of Economic Studies*, Vol. XXXIV, No. 97, 1967.
3. Gilbert, R. J., "Optimal Depletion of an Uncertain Stock," Technical Report No. 207, Institute for Mathematical Studies in the Social Sciences, Stanford University, 1976.
4. Heal, G. M., *The Theory of Economic Planning*, Amsterdam: North Holland Publishing Company, 1973.
5. Kemp, M., "How to Eat a Cake of Unknown Size," Chapter 23 in M. Kemp, *Three*

Topics in the Theory of International Trade, Amsterdam: North Holland Publishing Company, 1976.

6. Loury, G. C., "The Optimum Exploitation of an Unknown Reserve," Mimeo, Northwestern University. 1976.
7. Rothschild, M., and J. E. Stiglitz, "Increasing Risk I: A Definition and II: Its Consequences," *Journal of Economic Theory*, 2 (1970), No. 3 and 3, No. 1.

SEARCH STRATEGIES AND PRIVATE INCENTIVES FOR RESOURCE EXPLORATION

Richard J. Gilbert, UNIVERSITY OF CALIFORNIA, BERKELEY

I. INTRODUCTION

The search for deposits of nonrenewable resources serves two different economic functions. Exploration is a means to expand the capacity to extract a natural resource. Production capacity can be increased by more intensive development of known deposits or by identifying and developing new deposits. Exploration also serves to produce information about the resource stock, which includes information about (a) the amount of the total exploitable stock of a resource, (b) the cost of extraction of different deposits (including such characteristics as the level of impurities in different ores), and (c) the location of different deposits.

Advances in the Economics of Energy and Resources, Vol. 2, pp. 149–169.
Copyright © 1979 by JAI Press, Inc.
All rights of reproduction in any form reserved.
ISBN: 0-89232-079-6

Information about the location of deposits avoids or reduces the incidence of "dry" holes. The value of information about total resource stocks is more subtle. Improved knowledge of the total stocks of resources permits choice of current consumption rates that in an expected value sense allows a better balance between the welfare of present and future generations. Without this information, there may be a greater risk of an unexpected short-fall, or surplus, in future resource supply.

For some fossil fuels, such as coal and bituminous shales, and for several minerals the total crustal abundance may be considered virtually inexhaustible. However, crustal abundance is a very poor measure of resource supply because it aggregates important economic characteristics. For convenience, we lump these characteristics that affect the market value of a resource under a general heading of extraction cost, but they include such diverse factors as level of impurities (e.g., the sulfur content of coal and oil), proximity to markets, environmental impacts associated with extraction, processing and transportation, etc. In most cases where the total stock of a resource is very large, the total amount of high-quality ore may be small enough to be a constraint on the rate of economic growth.

This paper is an attempt to characterize private investment in resource exploration. The market incentive for exploration is clearly the value of the resources that may be discovered. The market offers no *direct* compensation for the information produced in the process of exploration. Section 2 is a simplified model of decentralized search for resource deposits, and serves as an introduction to the problem of the private allocation of exploration effort. A useful conclusion from this exercise is that exploration is equivalent to extraction under conditions of certainty if the probability distributions of discoveries at different locations are known and unaffected by search. Unexplored areas can be characterized by a certainty-equivalent extraction cost, and an efficient depletion policy would call for exploitation of areas in order of increasing cost.

The results of section 2 are similar to the implications of models of depletion under certainty because the process of search in this example provides no information that may revise expectations of future returns from exploration.[1] The remaining sections of this paper explicitly consider the implications of information in private exploration decisions. The first question we ask is the very basic one of existence of equilibrium when information that affects the returns from drilling may be produced. Section 3 shows the existence of an equilibrium where deposits that are identical *ex ante* are explored at different intensities.

The indirect compensation provided by the market need not generate the socially desirable level of investment in exploration. The market allocation may differ from the social optimum because(a) exploration is risky, (b) exploration may benefit people who are not required to pay for the information, and (c) the private speculative value of exploration information may differ from its social value.

In section 2 we argue that exploration is not a particularly risky activity because it is a diversifiable risk. Section 3 briefly discusses the speculative value of exploration information and neighborhood effects. The discussion here is limited to the effects of speculation such as occurs in competitive lease auctions, and of local information externalities on the private rate, timing and location of exploration. A more detailed contrast of the social and private value of exploration information is contained in Gilbert (8).

2. EXPLORATION AND SEQUENTIAL SEARCH

2.1 A Simple Model

We begin with several assumptions.

A.1 *Individual prospectors are risk-neutral.* From a societal perspective, the risk in resource exploration is the uncertainty in total discoveries. Any one well may be a failure, but with many thousands of wells, the total risk to society should be small if the returns from each well are uncorrelated. However, the private costs of drilling wells and sinking mine shafts can be very high, and the probability that any one venture is successful may be low. In spite of the high costs and low probability of success, it is likely that the private risk in resource exploration is *not* particularly great relative to the risk present in other sectors of the economy. The reason is that exploration is a diversifiable risk. It is not the return from any single venture that determines the financial risk, but rather the total return from a portfolio of ventures. A typical firm will drill several wells in a given year. Provided the manager of that firm does not make systematic errors in assessing the cost and prospective returns from drilling, then the disappointments at some locations should be at least partially offset by success at other locations. The opportunities for diversification are not limited to a company's own drilling activities. Oil prospectors typically issue shares in their drilling projects and buy shares in other projects, thereby spreading the risk per dollar over a large number of projects. The joint arrangements of drilling ventures can take on many forms that permit a better allocation of risk-bearing.

A test for risk is the market compensation for those who bear the cost. If exploration were very risky, one would expect to see a high average rate of return. The empirical evidence is not what one would expect for a risky industry. The after-tax rate of return for the eight largest integrated petroleum firms has averaged about 11 percent in the period 1960 to 1970, compared to about 10.5 percent for all manufacturing.[2] For these reasons, we argue that the assumption of risk-neutrality in the exploration decision is not particularly unreasonable.

A.2 *The marginal extraction cost, CE, of each deposit is constant.* The cost is defined to include all relevant factors that influence the production cost of the resource. The assumption of constant marginal costs considerably simplifies the analysis. [See Weitzman (32) for a model of depletion under conditions of certainty with arbitrary costs.]

The mechanics of the model may be more easily described if we assume that there are only a finite number of deposit grades, indexed by $CE_j, j = 1, \ldots, n$, where

$$CE_1 < CE_2 < \ldots < CE_n.$$

This is not a restrictive assumption since we are free to make the set of grades arbitrarily dense to approximate a continuum to any desired precision.

A.3 *Each location, or "tract," is characterized by a joint probability distribution on the size and extraction cost of the deposit at that location.* The sizes and extraction costs of deposits at different tracts are statistically independent. Within a "basin" the probability distribution is assumed identical for all tracts. There are m basins, indexed $i = 1, \ldots, m$.

A.4 *After the location is drilled (sampled), the size and extraction cost of the deposit is revealed.* There is no technology for the production of information about the characteristics of tracts before they are sampled.

Assumptions A.3 and A.4 are quite strong, and the consequences of relaxing these assumptions are examined in section 3. In particular, assumption A.3 implies that the expected proportion of new discoveries in basin i with extraction cost CE_k is the same at every location. If we denote this proportion by w_k^i, then

$$w_k^i = \frac{\bar{S}_k^i}{\sum_{j=1}^{n} \bar{S}_j^i}, \tag{1}$$

where $\bar{S}_j^i$ is the expected size of a deposit with extraction cost CE_j in basin i. The w_j^i, $j = 1, \ldots, n$, are independent of cumulative discoveries in the basin, at least until all tracts have been sampled (i.e., we allow the possibility of successive search in a basin). An alternative interpretation where the w_j^i depends on cumulative discoveries is discussed in section 2.2.

If the returns from drilling at different tracts are independent, constancy of the w_j^i follows from the definition of a basin as a region where the distribution of discoveries is the same at every location. The difficulty is that, in general, a discovery at one location will revise expectations of discoveries at neighboring tracts. The effects of such correlation on the allocation of search effort is the subject of section 3.4.

Assumption A.4 implies that exploration in this model is simply the decision to drill in an area where the return at each location is uncertain, but the probability distribution is known. This follows from the assumptions of independence of tracts and the absence of a technology to screen deposits prior to drilling. Some of the implications of exploration as a means to produce information are discussed in Section 3.

We require one additional assumption:

A.5. *There are no costs incurred in moving from one basin to another.* Assumption A.5 refers only to possible "set-up" costs associated with initiating search in a new area. If these costs were not negligible, the total amount of the resource that could be recovered from a basin would be a factor in the search decision. Given assumptions A.1–A.5, we shall show that the search problem can be transformed into a problem of depletion under conditions of certainty. Each unexplored basin can be characterized by a reservation cost, and the optimal strategy is to exploit the area with the lowest reservation cost. Furthermore, the market-determined sequence of exploration should be efficient.

We denote the total rate of discovery by d(t). The quantity d(t) represents the total amount of the resource found per unit time, aggregated over all extraction costs $CE_1, \ldots, CE_n$. Let CD^i be the expected marginal discovery cost in basin i. This is the expected change in exploration expenditures required to discover another unit of the resource in basin i, aggregated over all extraction costs. The cost CD^i reflects the probability that some locations may be "dry." We assume that marginal discovery costs in each basin are constant over time and independent of the rate of discovery. Changes in discovery costs due to depletion are discussed in section 2.2.

The stock in basin i of identified (proven) reserves of the resource with extraction cost CE_j is

$$R_j^i(t) = \int_0^t \left(w_j^i d(t) - q_j(t) \right) dt + R_j^i(0), \tag{2}$$

where $q_j(t)$ is the rate of extraction from deposits with extraction cost CE_j and $R_j^i(0)$ is the initial stock of identified reserves. The amount of the resource with extraction cost CE_j remaining to be discovered in basin i is

$$X_j^i(t) = X_j^i(0) - \int_0^t w_j^i d(t) dt. \tag{3}$$

The initial total stock of the resource is $X_j^i(0)$. The expected return from an optimal exploration program in basin i is

$$\Pi^i = \max_{d(t), q_j(t)} \left[\int_0^\infty \left\{ P(t) \sum_{j=1}^n q_j(t) - \sum_{j=1}^n CE_j q_j(t) - CD^i d(t) \right\} e^{-rt} dt \right], \tag{4a}$$

where P(t) is the expected price of the resource for time t and r is the market rate of interest (assumed constant). The control variables, d(t) and $q_j(t)$, are chosen subject to the constraints that

$$X_j^i(t) > 0 \text{ and } R_j^i(t) \geq 0, \tag{4b}$$

for $j = 1, \ldots, n$.

We can anticipate some of the results of this formulation. Exploration will identify deposits with a range of extraction costs which may be either produced or held in reserve for possible extraction at a later time. The decision to extract will depend only on the extraction cost of the deposit and the expected rate of change of prices. If, for any deposit with extraction cost CE_j, the expected change in rent yields a rate of return less than the rate of interest, the deposit will be produced immediately after discovery. If rent is appreciating faster than the rate of interest, the deposit will be held for possible extraction at a later time. This should be clear from the structure of the certainty problem.

There is a cut-off grade, CE_k^i, that minimizes the cost of search in basin i. Only deposits with extraction costs $CE_j \leq CE_k^i$ should be extracted immediately after discovery. The cut-off grade is determined by the identity

$$CD^i = \sum_{j=1}^k (CE_k^i - CE_j) w_j^i. \tag{5}$$

The left-hand side of equation (5) is the marginal cost of search. The right-hand side is the expected marginal reduction in the cost of extraction, relative to the cut-off grade CE_k^i, that results from search. For the cut-off grade the marginal benefit of search equals the marginal cost of search. If a deposit with $CE_j < CE_k^i$ is discovered, the expected saving from continued search is less than the cost of search and the deposit should be extracted. Conversely, for $CE_j > CE_k^i$, there is a positive net benefit from continued search and extraction from the deposit should be postponed. If there is no CE_j for which equation (5) is an identity, the cut-off grade is the highest cost for which the right-hand side of equation (5) is less than CD^i.

Equation (5) parallels the results in the theory of consumer search with random price offers and job search with random wage offers [see Rothschild (26), David (4) and McCall (20)]. The cut-off grade is equivalent to a reservation or acceptance cost. The expected cost of search in basin i when the cut-off grade is CE^i_k is

$$C^i(k) = \frac{1}{W^i(k)}\left[CD^i + \sum_{j=1}^{k} CE_j w^i_j\right], \tag{6}$$

where $W^i(k) = \sum_{j=1}^{k} w^i_j$.

Substituting equation (5) in equation (6) gives

$$C^i(k) = CE^i_k.$$

The cut-off grade equals the expected cost of search and extraction in basin i.

Of course, it is possible that

$$CD^i > \sum_{j=1}^{n} (CE_n - CE_j)w^i_j,$$

in which case the cost of search in the i-th basin exceeds the expected cost savings from additional search. All deposits would be extracted upon discovery and there would be no accumulation of deposits for possible extraction at future times. The effective production cost in the i-th basin would then be

$$\hat{C}^i = CD^i + \sum_{j=1}^{n} CE_j w^i_j = CD^i + \langle CE \rangle^i,$$

where $\langle CE \rangle^i$ is the average extraction cost of all deposits in the i-th basin.

Consider a continuum of deposit costs distributed over the interval $[\underline{CE}, \overline{CE}]$ according to the probability density function $w^i(CE)$. Writing the integral representation of equation (5) and integrating by parts gives

$$CD^i = \int_{\underline{CE}}^{CE^i_k} W^i(CE)d(CE), \tag{7}$$

where $W^i(CE)$ is the cumulative probability distribution of costs in basin i. Suppose equation (7) is satisfied for some $CE^i_k < \overline{CE}$ and consider the effect on the reservation cost of a mean-preserving spread in the distribution of deposit costs. By definition, if $W'(CE)$ is a mean-preserving spread of $W(CE)$, then

$$\int_0^{CE} W(v)dv < \int_0^{CE} W'(v)dv, \tag{8}$$

for all $CE < \overline{CE}$. [see Rothschild and Stiglitz (26).] For the distribution, $W'(CE)$, equality in (7) requires a lower CE^i_k. Thus basins that are more "risky" in the Rothschild-Stiglitz sense have *lower* reservation costs. The intuitive explanation of this result is simple enough. A more "risky" distribution has more weight in both the high-cost and low-cost tails of the distribution. However, the firm has the option of passing by poor deposits so that the interesting part of the distribution is only the low-cost tail.

Clearly, the optimal search rule is to exploit the area with the lowest reservation cost. This could be an area of proven reserves or an unexplored area. A market economy with a complete set of utures markets, or an economy where expectations are intertemporally consistent, would achieve precisely this result. If there are futures markets or if expectations are consistent, the resource price would increase over time so that the expected net rent for the area with the lowest extraction cost would rise at the market rate of interest. Firms operating in higher cost areas would experience capital gains by postponing search.

2.2 Depletion Effects

In the preceding analysis the distribution of deposits discovered in the process of exploring a basin and the marginal cost of exploration were independent of total discoveries in a basin (until cumulative discoveries equaled the total amount of a particular grade of the resource). This logically follows from assumptions A.3 and A.4. The returns from drilling at locations in a basin were assumed to be independent, identically distributed random variables, and the process of search did not provide information about future discoveries in a basin. The change in expected costs associated with depletion is not a continuous process in this model. Expected costs remain constant until every tract in a basin has been sampled, and then increase abruptly when a new basin is exploited (or the old basin may be searched again with a higher reservation cost). If there are many basins, the cost increase may be approximately smooth, but exploration is inherently a process of sequential search, moving from one area to another with a higher reservation cost.

Yet there are situations where one might expect search costs and the distribution of discoveries to change with cumulative discoveries in a basin. If large deposits cover several tracts, on average the larger deposits would be discovered early in the search of a basin and search costs may increase with cumulative discoveries.[3] Another possibility is that basins may be loosely defined to include tracts with different probability distributions or extraction costs. These situations violate the assumption that the returns from exploration are independent, identically distributed random variables, and some of the complications that arise when discoveries at different tracts are correlated are dealt with in section 3. Nonetheless, we may easily show that the previous results for the optimal sequence of exploration and extraction are unchanged if there exists a reservation cost of search which is a nondecreasing function of total discoveries. This would be the case if CD^i

increased or the w_j^i decreased with cumulative discoveries [cf. equation (5)]. Efficiency still calls for production from those deposits which have the lowest reservation cost, whether those deposits are in previously unexplored territories or in the stock of identified reserves.

In this case the derivation of competitive equilibrium price trajectories is more complicated, but the previous results for the optimal sequence of exploration and extraction are not changed. It is not difficult to show that efficiency still calls for production from those deposits which have the lowest current marginal factor cost, whether those deposits are in previously unexplored territories or in the stock of identified reserves.

To show this, assume that the reservation cost of search and extraction in basin i is less than the cost of depleting identified reserves or exploring any other basin. Let $C^k(t_0)$ be the current reservation cost of an alternative area and let $\lambda^k(t_0)$ be its associated scarcity rent at time t_0 discounted to the present. If current production from the k-th area is positive, efficiency requires

$$C^k(t_0) + \lambda^k(t_0)e^{rt_0} \leqq C^i(t_0) + \lambda^i(t_0)e^{rt_0}, \tag{8}$$

or

$$[\lambda^i(t_0) - \lambda^k(t_0)] \geqq [C^k(t_0) - C^i(t_0)]e^{-rt_0} > 0. \tag{9}$$

It is possible to write the Hamiltonian of the profit-maximization problem as

$$H = \left[P(t) \sum_{j=1}^{n} q_j(t) - \sum_{j=1}^{n} C^j(t) q_j(t) \right] e^{-rt} - \sum_{j=1}^{n} \lambda^j q_j(t), \tag{10}$$

where

$$\frac{d\lambda^j(t)}{dt} = \frac{-\partial H}{\partial X_j} = \frac{\partial C^j}{\partial X_j} q_j e^{-rt} = -\frac{dC^j}{dt} e^{-rt}. \tag{11}$$

Consequently,

$$\frac{d}{dt}\left[C^j(t) + \lambda^j(t)e^{rt} \right] = r\lambda^j e^{rt}. \tag{12}$$

Since $\lambda^i(t_0) > \lambda^k(t_0)$, then by equation (12), it follows that $\lambda^i(t) > \lambda^k(t) \geqq 0$ for all $t \geq t_0$. Thus production from the i-th basin is deferred indefinitely and its opportunity cost, λ^i must be zero. This is a contradiction, so we conclude that extraction is in the order of increasing reservation costs.

The argument fails if factor costs may decrease with cumulative discoveries or if there are costs in moving from one basin to another. In this event, it may be efficient to produce an area with factor costs higher than other areas, if production leads to significant reduction in future costs. For a discussion of this problem see Weitzman (33).

3. EXPLORATION AND THE PRODUCTION OF INFORMATION

The simple model assumed that the probability distributions of deposit costs for different basins were known before drilling began. The drilling of a well at a particular location revealed the cost of extraction at that location. The decision to extract the reservoir or postpone recovery could then be made given perfect knowledge of the extraction cost at that location and given the probability distributions of costs at other locations.

This is a considerable simplification of the exploration process in at least two important respects. First, a substantial amount of geological and geophysical work may precede the decision to drill at a particular location. Outlays on these information generating activities in recent years have been on the order of 20 percent of drilling costs in the United States.[4] The opportunity to produce information about deposits prior to drilling implies that the probability distributions of costs at different locations should not be taken as given and immutable. The production of information prior to drilling allows a partial screening of tracts that improves the expected returns at those locations where geological and geophysical indications are favorable.

A second important construct of the preceding analysis was the assumption that the returns from drilling at different locations were statistically independent. Success or failure of a particular well will lead to a revision of the probabilities of discovery at locations with similar geological characteristics. If the activities of one firm affect another, there is an externality and a potential distortion in the private rate, timing and location of exploration.

This section is organized in four parts. Before embarking on an analysis of the value of information, we consider in section 3.1 the basic question of existence of equilibrium when deposits can be screened. The model is an extension of the problem studied in the previous section, the difference being that exploration can change the probability distribution in a basin before drilling begins. Although still simple, the model illustrates problems that are peculiar to the production of information in a market economy.

Exploration information produced by one firm is often of value to other firms and individuals. If the information is public there may be positive externalities. There are two difference external effects. One has to do with the returns from drilling at tracts with similar characteristics (i.e., where the returns are correlated), which we call local information externalities. The other externality concerns information about the total resource stock. The current price of an exhaustible resource depends on knowledge of the total resource stock and future extraction costs. Firms have a direct incentive to produce information about the total resource stock only if they can profitably speculate on the resource price, and this is briefly discussed in section 3.2.

A competitive lease auction is an invitation for private speculation. By producing information about a tract that is offered for lease in a competitive auction, firms can submit a better-informed bid that may provide a larger margin for profit. Since land is often allocated among firms through a bidding process, section 3.3 examines incentives in an auction market that affect the rate, timing and location of exploration. The final part of this section, section 3.4, is a discussion of the allocative effects of local information externalities.

3.1. Exploration Information and Market Equilibrium

We begin this section with a discussion of market equilibrium when firms may invest in information prior to drilling. The assumption of independence of tracts will be retained in this analysis. Note that even if tracts are statistically independent there are pecuniary externalities among firms. The value of any one lease, and the expected return from exploration, may depend on information that has been produced about other properties. The value of any one lease depends on the current and expected market price. Current and future prices depend on information about extraction costs and the total resource stock, which depends on the exploration of other tracts. Put another way, information about the cost of a particular deposit is a public good if it affects the equilibrium price.[5]

A precise statement of the nature of market equilibrium with pre-drilling exploration can be made with the aid of a simple example. [A more detailed discussion of this problem is in Gilbert (10)]. Suppose there is only one area where the resource may be found, and the expected unit drilling cost at each tract in the area is CD. There are only two kinds of deposits; one has a constant marginal extraction cost, CE_l, while the other has a higher cost CE_h. Let π be the expected proportion of discoveries with extraction cost CE_l, in the absence of any pre-drilling information. We assume that

$$CD > \pi(CE_h - CE_l),$$

so that both types of deposits would be exploited if there were no exploration information which could be used to screen tracts before drilling. The expected total production cost would then be

$$\bar{C} = CD + \overline{CE},$$

where $\overline{CE} = \pi CE_l + (1 - \pi)CE_h$. In equilibrium, prices should rise over time so that

$$\frac{\dot{P}_t}{P_t - \bar{C}} = r.$$

Now suppose a technology becomes available that allows perfect identification of low-cost deposits before drilling at a unit cost, e. Certainly if e is sufficiently small, at prevailing prices it would be profitable to screen the low-cost from the high-cost deposits. Indeed, taking prices as given, all tracts would be screened. Yet, if the high-cost deposits are exploited after the low-cost deposits are depleted, we can show that complete screening cannot be an equilibrium with perfect foresight. Suppose the low-cost deposits are extracted during the time interval T_0 to T_1, and the high-cost deposits are used from T_1 to T_2. In equilibrium, a producer must be indifferent with respect to extracting the low-cost deposits at any time between T_0 and T_1, and, similarly, with respect to extracting the remaining high-cost deposits between T_1 and T_2. Figure 1 illustrates an expected price trajectory with this property. Compare the returns from screening to the returns from production at time T_1 without pre-drilling exploration. In the latter case the expected cost is $\bar{C}$, and this is equivalent to producing an expected proportion, π, of low-cost deposits immediately before time T_1 and producing the remaining high-cost deposits immediately after time T_1. That is, at time T_1, the expected revenue earned when deposits are screened is equal to the expected revenue without screening. However, profits are lower with pre-drilling exploration by the amount, e. The original screening choice could not be an equilibrium with perfect foresight, since after market prices adjust to reflect the new information, producers would wish to change their original decisions.

There is an equilibrium market level of screening in this example, and it requires that pre-drilling exploration be limited to a fraction of the total area. An equilibrium price path is shown in Figure 2. In the interval T_0 to T_1', those deposits that have been screened and are low-cost are extracted.

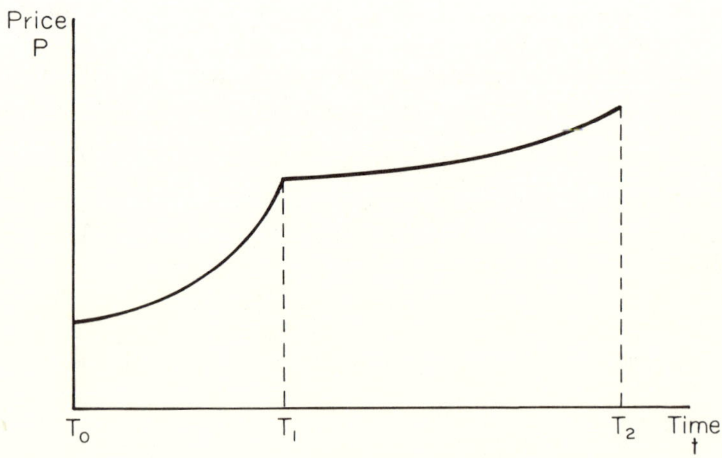

Figure 1.

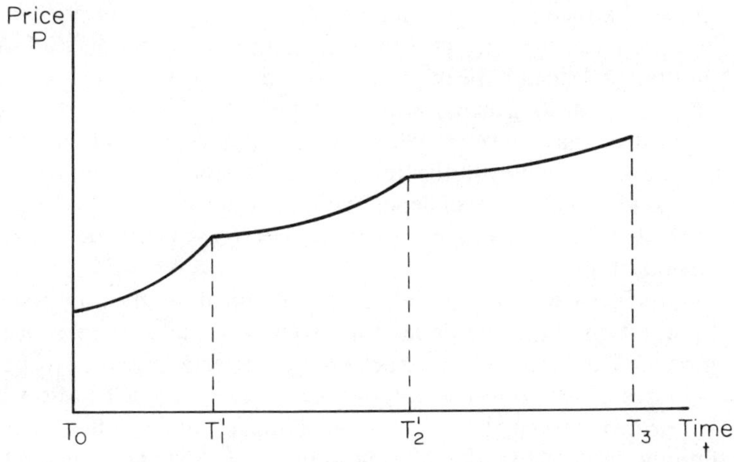

Figure 2.

From T'_1 to T'_2, production is from unexplored deposits, while the remaining high-cost deposits are produced from T'_2 to T_3.

It is not difficult to show that the equilibrium is efficient; yet it is somewhat surprising because locations that are identical *ex ante* are not matched with identical production technologies. This result clearly hinges on the assumption that pre-drilling exploration does not eliminate dry holes, but instead separates tracts into one category that should be exploited "now" and another for use "later." One may question the realism of this example, but it serves to illustrate the nature of market equilibrium with endogenous information.[6]

3.2 Exploration and Price Speculation

In the preceding example, investment in exploration at a market equilibrium was efficient but this need not be the case. Consider a slight modification of this example. Again there is one area where the resource may be found. At each location there may be a deposit of size S with extraction cost CE, but the total proportion of tracts that contain the resource and hence the total stock of the resource is uncertain. In general, uncertainty in the size of the total resource stock will be reflected in the spot price of the resource. Relative to a known total stock, an uncertain stock with the same mean should have a higher equilibrium spot price [see Gilbert (8), Loury (19), and Heal (14)]. The difference can be thought of as a risk premium, and information about the true size of the stock has social value.

Do private firms have an incentive to produce information about the total resource stock? Unless firms speculate on the resource price, the answer is No. Information about the total stock, in this example, has no effect on the cost of extraction; i.e., the information affects only the price of the resource. Of course, firms may have an indirect incentive to produce information about the total stock if the information lowers expected extraction costs. For example, pre-drilling exploration that permits a partial screening of promising tracts may lower expected extraction costs, and, indirectly, provide a better estimate of the total resource stock. Alternatively, the cost of extraction and the size of deposits could be correlated, and the search for low-cost deposits may provide (again indirectly) information about the total resource stock.

Speculation is one way to profit from information about the total resource stock. Information about the price of a resource will affect asset values. If exploration indicates that the total stock exceeds expectations, the prospector can buy shares of assets whose values will increase and sell those whose value will decrease when the information is made public.[7] Of course, the opportunity to trade before the information is public may be quite limited.[8] If a firm is large enough, e.g., if the firm is a resource monopoly, the change in the firm's own value may be sufficient to induce exploration of the total stock [this point is made by Stiglitz (29) in the context of securities markets].

Information about future extraction costs should also affect the spot price of a resource and the current rate of depletion. There are vast deposits of bituminous shales that can produce an almost perfect substitute for natural crude oil. The production cost of shale oil is uncertain, and information that these resources could (or could not) be extracted at, say, \$10 per barrel should have an effect on the current price and rate of extraction of oil. A price-taking firm would not invest in the exploration of oil shale or other high-cost deposits until lower-cost deposits have been depleted, since this strategy minimizes the discounted present-value cost of exploration and extraction.

The right to produce oil shale or other high-cost deposits may have considerable value even if the deposits cannot be profitably exploited for many years. A deposit has a positive present-value if it can be profitably exploited (due to, e.g., increased prices or advances in technology) at some future date. If high cost deposits are offered in a competitive auction, firms may explore in order to maximize the expected profit from bidding. An auction market may provide incentives for firms to explore areas that, in the absence of the auction, would not be explored for some time. In the next section we examine the implications of competitive bidding for the private allocation of exploration effort.

3.3 Exploration and Competitive Bidding

A lease auction market is an invitation to engage in speculation. Since most mineral rights in this country are eventually awarded to the highest bidder, it is important to assess the effect of competitive bidding markets on the rate, timing, and location of exploration.

The expected profits of participants in a lease auction depend on each bidder's knowledge of the value of the lease and the strategies of other bidders. There is much to be learned about the operation of complex auction markets, but there is a strong presumption that a participant with superior information may earn (monopoly) profits at the expense of less-informed bidders. Gaskins and Teisberg (7) cite a particular example (offered by Robert Wilson) where this is the case. Teisberg (32) has simulated an auction in which participants bid a multiple of their estimates of the value of the lease. His results show that profits decrease as the number of bidders increase, and that better information increases expected profit. While the bidding strategy assumed by Teisberg is not necessarily individually optimal, it conforms to reports of actual bidding strategies in lease auctions [See Capen, *et al.* (3) and Dougherty and Nozaki (5)].

In order to focus on the allocative effects of competitive bidding, we assume that exploration is limited to the search for low-cost deposits. We ignore neighborhood effects and the value of information about the total stock, and assume that the market is in equilibrium. We also ignore regulatory and fiscal measures, such as diligence clauses and distinctive tax allowances, that may distort the market-determined pattern of exploration. Under these conditions, there is no wedge between the private and social value of exploration. If a firm owns the mineral rights to a particular tract, its exploration activities should be efficient from the perspective of maximizing the social product. However, if a firm competes for mineral rights in a lease auction, exploration information may increase the expected profit from participation in the auction and the private investment in information need not be socially optimal.

First consider the effect of competitive bidding on the timing of exploration. Let $V(t)$ be the expected value of a particular tract at time t if there is no pre-drilling exploration, and let $V(t|E)$ be its expected value conditional on an investment, E, in pre-drilling exploration. It is reasonable to assume that $V(t|E)$ is non-decreasing in E and

$$V(t|E) > V(t)$$

for $E \geq \hat{E}$. That is, there may be a threshold value, $\hat{E}$, below which pre-drilling information is useless; but for $E > \hat{E}$ we assume that the value of information is strictly positive.

The optimal date and level of exploration of the tract would maximize its net discounted present value

$$e^{-rt}(V(t|E) - E),$$

with respect to t and E. We assume that the optimal date, T^*, and level of exploration, E^*, are unique, and $E^* \geq \hat{E}$.

Suppose the lease is offered in a competitive auction at date T*. Will the market-determined level of exploration be E*? A definitive answer to this question requires a careful study of bidding behavior, but preliminary results support a negative conclusion. Note that there may well be redundant investment in exploration, as each firm seeks information to improve its expected profit from the auction. Gaskins and Teisberg (7) observe that exploration prior to a lease sale could be socially productive in assigning a lease to the firm that, as the result of its superior performance, places the highest value on the lease. However, the bidding process performs this matching function only if firms differ in abilities and not merely in their information. This information could be made public, and if firms have equal abilities and the same priorities they should calculate the same expected value for the lease.[9]

Furthermore, the simulation results of Teisberg (32) suggest that the private return to information in a competitive lease auction is proportional to the expected value of the lease. Yet the appropriate criterion for exploration is its marginal value. It is at least conceivable that some lucrative tracts may require little or no exploration, but the lure of monopoly profits would motivate pre-drilling exploration in the lease auction.[10]

Suppose a lease auction takes place at a date τ before T*. The present value of the tract may be positive even if the average extraction cost exceeds the current spot price, provided the price is expected to exceed average cost at some future date.[11] If τ is sufficiently close to T*, the present value of the tract may be large enough to stimulate excessive pre-bidding exploration. However, for $\tau \ll T^*$, the expected present value of the tract may be less than $\hat{E}$, the threshold level of investment in exploration necessary to provide any useful information about the tract. If the lease auction sufficiently antedates T*, and if there is a threshold level of exploration, pre-bidding exploration may not be profitable. The lease would be awarded to the highest bidder, and, in the absence of external constraints, the lessee would postpone exploration until the optimal date T*.

3.4 Local Information Externalities

If the returns from drilling at locations owned by different firms are not independent, exploration by any one firm confers external benefits on other firms. We call these external benefits local to differentiate them from externalities associated with the total resource stock. If firms do not coordinate their exploration activities, local information externalities will distort the private allocation of exploration effort relative to the socially optimal allocation.

The nature of the distortion may be illustrated by assuming there are only two tracts. Exploration is the decision to drill a well at cost E. If successful, the well discovers a pool of oil of value V, and nothing otherwise. With no information, the probability of discovery at either tract is π. If exploration at one tract is successful, the conditional probability of success at the other tract is $\pi_S > \pi$, and if exploration is a failure, the conditional probability is $\pi_F < \pi$.[12] A joint exploration strategy would be sequential with expected value (ignoring time discounting) of

$$R_{total} = (\pi V - E) + \pi\{\pi_S V - E\} + (1 - \pi)\max\{\pi_F V - E, 0\}.$$

The expected value of independent exploration at each tract is just

$$R_i = \pi V - E.$$

Since $\pi_S > \pi$, it is clearly possible that $R_i < 0$, but

$$R_{total} = (1 + \pi)R_i + \pi(\pi_S - \pi)V > 0.$$

If the tracts are owned by different firms and if the results of exploration by one firm are not available to the other, it is possible that neither firm would explore even though joint exploration would be profitable. Alternatively, with no exchange of information there may be too much exploration. If the first well does not discover oil, and if

$$\pi_F V < E,$$

the second well should not be drilled; but acting independently both firms may drill wells.

Conversely, if the results of exploration are public, the first firm to explore makes an expected profit of

$$R_1 = \pi V - E,$$

while the second makes

$$R_2 = \pi(\pi_S V - E) + (1 - \pi)\max(\pi_F V - E, 0).$$

If $\pi_F V - E < 0$, R_2 exceeds R_1 and it pays to wait for thc other firm to explore. Of course, postponement of exploration benefits neither firm and there is an inducement to cooperate in a joint exploration strategy.

In this simple example there are only two possible locations where the resource may be discovered. Extending the problem to consider the choice of drilling location in a large area, we note that the optimal pattern of drilling should balance the information gained by drilling in unexplored areas against the increased probability of discovery by drilling close to known successful wells. If the results of drilling are private, a firm's choice of location will be predicated on its own success and there will be too little exploitation of information produced by others. If the results of drilling are public, Stiglitz (30) has shown that firms will tend to cluster in the neighborhood of successful wells and there would be too little drilling in relatively unexplored areas.

4. CONCLUDING REMARKS

The search model in Section 2 described the function of exploration as a means to expand production capacity through search, while the remainder of this paper has centered on the information value of exploration. In many areas the results of a drilling program may be uncertain, but as a consequence of prior exploration the distribution of returns may be known with some precision. The information value of search is limited in these areas and we would expect the model to be a reasonable description of the exploration process. Several conclusions emerge from this analysis.

1. The deposits which have the lowest production cost at any particular moment of time will be exploited. These deposits may be as yet undiscovered if the expected cost of exploration and extraction is less than the extraction cost of identified deposits. Prices will adjust so that producers will earn the market rate of return on the best deposits.

The formulation presented in this paper contradicts the causal relationship implied by Adelman (1) who states (page 74), "...Exploration is needed to prevent an otherwise inevitable rise in developing-operating costs. The oil operator seeks new reserves to keep his costs down."

The model in Pindyck (23) links extraction costs to the stock of proven reserves and formalizes this hypothesis. In the model discussed in this paper, reserves are identified during the process of search, but the level of reserves per se provides no incentive for exploration. Exploration continues as long as the cost of discovering and extracting new deposits is less than the cost of extracting known deposits, and the notion of maintaining an inventory to keep costs down does not arise.

2. In the process of exploration, extraction from some deposits that are discovered may be postponed. Although the search rule maximizes profits, it is not surprising that lease provisions may cause the tenant to choose an inefficient extraction policy. For example, the common diligence clause requires that production begin by a specified time and be sustained to renew the lease. However, the efficient policy could call for postponement of development. Given the alternative of losing the lease, the tenant would choose to extract from the deposit provided net rent is positive.

3. When stocks of a particular quality are exhausted, effective production costs increase. At this point, the optimal production choice may call for exploration and extraction in a different basin, or extraction from the stock of identified reserves. It is clear, then, that the optimal depletion policy may entail the recurrence of intervals of search and extraction from identified reserves, without any regular pattern. The sequence depends on the distribution of exploration and extraction costs in different basins.

Exploration generally produces information about the distribution of returns from search and the characteristics of the total resource stock.

The market may fail to produce the correct amount of information if exploration has external benefits. The externalities include information about neighboring deposits and information about the total amount and extraction cost of the resource. If the returns from drilling at different tracts are correlated, firms may not explore areas that should be explored.[13] There may be redundant exploration if information is private and a tendency to cluster drilling where information is public.

A lease auction encourages exploration as a means to compete in the bidding market. The exploration may provide valuable information about the resource stock, but the investment in information may be redundant and excessive.

FOOTNOTES

I am grateful to Partha Dasgupta, Joseph Stiglitz, and James Sweeney for many stimulating discussions and valuable observations.

1. Several authors have discussed the theory of extraction under conditions of certainty; see, e.g. Hotelling (17), Herfindahl (15), Nordhaus (21) Solow (28), and Sweeney (31).
2. These data are from the "Preliminary Federal Trade Commission Staff Report on its Investigation of the Petroleum Industry" (24).
3. Kaufman (18) gives an interesting statistical analysis of sampling proportional to size from a finite population.
4. See "Capital Investment in the World Petroleum Industry," The Chase Manhattan Bank, published annually.
5. A number of authors have discussed the public good nature of information in a general equilibrium context, e.g., Green (11), Grossman (12), Grossman and Stiglitz (13), and Radner (25).
6. Taking prices as given, the profit-maximizing investment in exploration need not be efficient, but this is conceptually no different from the problem of the stability of adjustment toward equilibrium in competitive markets under conditions of certainty.
7. The private return to speculation and its social value may be very different [see, e.g., Hirshleifer (16) and Stiglitz (29)].
8. Indeed, the change in asset prices that accompanies trade may be sufficient to reveal the traders' information [see Green (11) and Grossman (12)].
9. They may have different priors that result from previous exclusion of information. For an interesting discussion of information and agreement, see Aumann (2).
10. The possibility that market incentives may lead to too much exploration was observed by Gaffney (6).
11. This assumes the lease can be held dormant until any future date, which would not be the case if a diligence clause is binding.
12. This model is similar to the example presented in Stiglitz (30).
13. Peterson (22) also noted this result.

REFERENCES

1. Adelman, M., *The World Petroleum Market*. Johns Hopkins University Press, Baltimore, 1972.
2. Aumann, R., "Agreeing to Disagree," Stanford University mimeo, 1976.
3. Capen, E., R. Clapp, and W. Campbell, "Competitive Bidding in High Risk Situa-

tions," *Journal of Petroleum Technology* (June 1971), 641–653.

4. David, P., "Fortune, Risk and the Microeconomics of Migration," in David, P., and M. Reder (eds.), *Nations and Households in Economic Growth.* Academic Press, New York, 1974.
5. Dougherty, E., and M. Nozaki, "Determining Optimum Bid Fractions," *Journal of Petroleum Technology,* (March 1975), 349–356.
6. Gaffney, M., "Editor's Conclusion," in Gaffney, M. (ed.), *Extractive Resources and Taxation.* University of Wisconsin Press, Madison, 1967.
7. Gaskins, D., and T. Teisberg, "An Economic Analysis of Presale Exploration in Oil and Gas Lease Sales," in Masson, R., and P. Qualls (eds.), *Essays in Industrial Organization in Honour of Joe S. Bain.* Ballinger Press, Cambridge, Mass., 1976.
8. Gilbert, R., "Optimal Depletion of an Uncertain Stock," *Review of Economic Studies,* (October 1978), forthcoming.
9. ———, "The Social and Private Value of Exploration Information," University of California, Berkeley Working Paper No. 91, 1977.
10. ———, "Resource Extraction with Differential Information," *American Economic Review* (February 1977), 250–254.
11. Green, J., "Information, Efficiency and Equilibrium," Harvard Institute of Economic Research, No. 284, 1973.
12. Grossman, S., "The Existence of Futures Markets, Noisy Rational Expectations and Informational Externalities," Stanford University, Calf., IMSSS, No. 182, 1975.
13. ———, and J. Stiglitz, "Information and Competitive Price Systems," *American Economic Review* (May 1976), 246–253.
14. Heal, G., "Uncertainty and the Optimal Supply Policy for an Exhaustible Resource," University of Sussex mimeo, 1977.
15. Herfindahl, O., "Depletion and Economic Theory," in M. Gaffney (ed.), *Extractive Resources and Taxation.* University of Wisconsin Press, Madison, 1967.
16. Hirshleifer, J., "The Private and Social Value of Information and the Reward to Inventive Activity," *American Economic Review* (September 1971), 561–574.
17. Hotelling, H., "The Economics of Exhaustible Resources," *Journal of Political Economy,* 39 (1931), 137–175.
18. Kaufman, G., "Statistical Methods for Predicting the Number and Size Distribution of Undiscovered Hydrocarbon Deposits," presented at the American Association of Petroleum Geologists Research Conference, Palo Alto, California, August 1974.
19. Loury, G., "The Optimal Exploitation of an Unknown Reserve," Northwestern University mimeo, 1976.
20. McCall, J., "Economics of Information and Job Search," *Quarterly Journal of Economics,* 84 (February 1965), 113–126.
21. Nordhaus, W., "The Allocation of Energy Resources," *Brookings Papers,* 3 (1973), 529–570.
22. Peterson, F., "Two Externalities in Petroleum Exploration," in Brannon, G. (ed.), *Studies in Energy Tax Policy.* Ballinger Press, Cambridge, Mass., 1975.
23. Pindyck, R., "Optimal Exploration and Production of a Nonrenewable Resource," MIT Energy Laboratory Working Paper 77–013 WP (May 1977).
24. "Preliminary Federal Trade Commission Staff Report on Its Investigation of the Petroleum Industry," U.S. Senate, Committee on Interior and Insular Affairs, 1973.
25. Radner, R., "Problems in the Theory of Markets Under Uncertainty," *American Economic Review,* 60 (1970), 454–460.
26. Rothschild, M., "Models of Market Organization with Imperfect Information," *Journal of Political Economy,* 81 (November-December 1973), pp. 1283–1308.
27. ———, and J. Stiglitz, "Increasing Risk I, A Definition," *Journal of Economic Theory,* 2 (1970), 225–243.

28. Solow, R., "The Economics of Resources or the Resources of Economics," *American Economic Review* (May 1974).
29. Stiglitz, J., "Perfect and Imperfect Capital Markets," presented at the New Orleans Meetings of the Econometric Society, 1971.
30. ———, "The Efficiency of Market Prices in the Long-Run Allocation of Oil," in Brannon, G. (ed.), *Studies in Energy Tax Policy*. Ballinger Press, Cambridge, Mass., 1975.
31. Sweeney, J., "Economics of Depletable Resources: Market Forces and Intertemporal Bias," *Review of Economic Studies*, Vol. XLIV, No. 136 (February 1977), 125–142.
32. Teisberg, T., "A Model of Competitive Bidding with Application to Federal Oil and Gas Leasing," University of California, Berkeley, mimeo (1976).
33. Weitzman, M., "The Optimal Development of Resource Pools," *Journal of Economic Theory*, 12 (1976), 351–364.

INCREASING EXTRACTION COSTS AND RESOURCE PRICES

Donald A. Hanson, SOUTHERN METHODIST UNIVERSITY

I. INTRODUCTION

It seems likely that cost-reducing technical change in the resource extraction sector will not be able to keep pace with probable cost-increasing factors such as a declining success ratio in exploration, smaller discoveries, more remote locations and deeper mines or reservoirs. Regarding this situation, Geoffrey Heal (3) points out that as exhaustible resources are depleted, the remaining stocks may not command large scarcity rents, if extraction costs become very large.

The theory of exhaustible resources, as presented in the initial papers (e.g., Hotelling (5), Herfindahl (4), and Solow (8)) did not deal in detail with the

Advances in the Economics of Energy and Resources, Vol. 2, pp. 171–186.
Copyright © 1979 by JAI Press, Inc.
All rights of reproduction in any form reserved.
ISBN: 0-89232-079-6

increasing extraction costs associated with resource depletion. Two recent papers by Solow and Wan (7) and by Heal (3), have focused on this feature in an economic growth context. They found that the shadow price of the cost grade of the resource used currently satisfies a somewhat unintuitive differential equation. Here we attempt to provide additional insight into this equation by deriving it in a very simple efficiency context. It is also shown that the efficiency price of the extracted resource satisfies the same simple relationship as in the conventional theory with constant extraction costs.

The theory of exhaustible resources deals with characterizing efficient resource depletion paths and market equilibrium paths and the corresponding resource prices. Resource i with extraction cost b^i has an economic value, if b^i is sufficiently small, since it is available in a finite quantity and nonrenewable. Denote the present value by μ^i. The basic result from the theory of exhaustible resources is that in a world of certainty, the value of a resource must grow at a rate equal to the rate of return to capital r. Thus the value of resource i at t is $\mu^i e^{rt}$, provided r is constant. That is, the capital gain associated with holding the resource stock must equal the return on other assets [see Solow (8)].

If at time t, it is efficient to be using resource stock i, then the efficiency price of the extracted resource must be equal to

$$p_t = \mu^i e^{rt} + b^i. \tag{1}$$

Further, if it is efficient to use stock i on an interval (t_{i-1}, t_i), then from equation (1) $\dot{p}_t$ is given by

$$\dot{p}_t = r(p_t - b^i) \tag{2}$$

or

$$\frac{\dot{p}_t}{p_t} = r\left(\frac{p_t - b^i}{p_t}\right)$$

for all $t \varepsilon (t_{i-1}, t_i)$. That is, the growth rate of the extracted resource price must be some fraction of the rate of return to capital. This fraction is the ratio of the price of the unextracted resource to the price of the extracted resource. Here it will be shown that equation (2) holds even if extraction costs are increasing with time.

Suppose that it will eventually be economical to use all resources with extraction costs less than $\bar{b}$. Further, suppose that there are no economic gains associated with increasing the stock of resources with extraction cost b or greater. Then $\mu(\bar{b}) = 0$. Here it is shown that the present value of resources with extraction cost $0 \leq b \leq \bar{b}$ is given by

$$\mu(b) = \int_b^b e^{-rt_\beta} d\beta \tag{3}$$

where if the extraction cost is continuous and increasing on an interval, t_b is the extraction time of the resource with extraction cost b, and if the extraction cost is the constant b on an interval, t_b is the depletion time of that stock, assuming resources with least extraction costs are used first. Equation (3) gives a simple characterization of the present shadow prices associated with resources of varying extraction costs. It is easily seen that $\mu(b)$ is decreasing and convex. Hence, the present value of an additional unit of resource decreases more than proportionately as its extraction cost increases. A resource with a higher extraction cost has a lower present value not only because of the greater cost which will be incurred, but also because it is efficient to postpone its use further into the future.

Suppose that b is continuous and increasing on an interval. Analogous to equation (1), for time t_b when resource b is used

$$p_{t_b} = \mu(b)e^{rt_b} + b \tag{4}$$

from which

$$\dot{p}_{t_b} = r\mu(b)e^{rt_b} + \mu'(b)e^{rt_b}\frac{db}{dt} + \frac{db}{dt}. \tag{5}$$

But $\mu'(b)$ from equation (3) is $-e^{-rt_b}$, leaving only the first term in equation (5). Substituting $\mu(b)$ from equation (4) into (5) implies equation (2) still holds.

Equation (2) is an important relationship. It states that in this kind of theory, a combination of stable prices and significant resource scarcity rents are not efficient in a world with $r > 0$. Therefore, those who would argue that there should be a significant scarcity rent on an exhaustible resource and would also argue that its price should not increase as rapidly as in equation (2), must base their argument on a more complex analysis than is captured in the conventional theory of exhaustible resources.

Further, by integrating equation (2), it is immediately seen that the average resource value or shadow price on an interval $[0, T]$ is $(P_T - P_0)/rT$. Suppose resource flow R_t is perfectly inelastic and that T is the depletion time. Then P_T is fixed by the cost of a substitute. Typically $P_0 \ll P_T$. In this case the average resource shadow price is approximately P_T/rT. Hence it is independent of the distribution of resources among cost grades as long as the total resource stock remains fixed. With a constant extraction cost, high resource shadow prices tend to be concentrated immediately before the depletion time. However, with increasing extraction costs these rents may be spread out more evenly over the entire interval or perhaps concentrated on subintervals other than near the depletion time. For the typical case of $P_0 \ll P_T$, the average resource shadow price is essentially independent of the cost path b_t.

Section II derives the basic results in the simplest of efficiency problems:

resource demand is given exogenously and the planning horizon is fixed. The only question is how to sequence extraction from the different resource cost grades to minimize the present value of future extraction costs. Yet the μ(b)'s, which can be interpreted as the marginal present cost savings of an additional unit of resource with extraction cost b, come out of this problem. Section III analyzes the efficient growth problem, similar to references (7) and (3), and shows that the same characterization of resource efficiency prices holds when resource demand and the interest rate are determined endogenously. Section IV shows that the price path, equation (2), is a competitive equilibrium with perfect foresight even when extraction costs are continuous and increasing. Section V seeks to summarize the results in an intuitive manner and presents further conclusions.

2. EFFICIENCY AND EXTRACTION COST

The development of the individual mine or reservoir is not considered here. Instead it is assumed that the unextracted resources can be aggregated and classified into groups described by well-defined extraction costs. Further, it is assumed that costs are independent of the rate of extraction, there is no cost-reducing technical progress, and that present extraction does not provide new information affecting future extraction costs. The amount of resource in each extraction cost group is known initially. The analysis neglects the locational aspects of the problem and the resulting structure of the costs to deliver the resource to different destinations. Once extracted all resources are identical.

Let $\bar{X}^i$ be the amount of resource i with extraction cost b^i, and let $i = 1,3$ with $b^1 < b^2 < b^3$. The simplest extraction efficiency problem will be considered: for any exogenous path of total resource flow $\bar{R}_t$ on $[0,T]$, what is the set of paths R^i_t, $i = 1,3$, on $[0,T]$ which minimize the present value of future extraction costs. Specifically, minimize $\mathscr{E}$ where

$$\mathscr{C} = \int_0^t \left[\sum_{i=1}^{3} b^i R^i_\tau \right] e^{-r\tau} d_\tau \tag{6}$$

$$\sum_{i=1}^{3} R^i_t = \bar{R}_t \tag{7}$$

and for $i = 1,3$

$$\dot{X}^i_t = R^i_t \tag{8}$$

$$X^i_T = \bar{X}^i \tag{9}$$

$$R^i_t \geq 0 \tag{10}$$

where r is the discount rate assumed to be constant and X^i_t is the cumulative amount of resource i extracted by time t. Finally, it is assumed that the highest cost resource is needed but that it is sufficiently abundant so that

$$\bar{X}^1 + \bar{X}^2 < \int_0^t \bar{R}_\tau d\tau < \bar{X}^1 + \bar{X}^2 + \bar{X}^3.$$

The solution to this problem is well known (4, 8): use the least costly resource first until it is depleted, then use the next least costly resource until it is depleted, and finally use the most costly resource for the remainder of the interval.

The proof goes as follows: let $\tilde{R}_t^1$, $\tilde{R}_t^2$, $\tilde{R}_t^3$ be paths on (0,T) satisfying (7) and (10) such that for $j > i$ resource j is used on an interval before resource i is depleted. Is such a set of paths a candidate for an efficient solution? The answer is no, since a small perturbation of the paths can always be found which decreases $\mathscr{C}$. The perturbation involves shifting the lower cost resource backward in time and the higher cost resource forward. Note that there will always exist an $a > 0$ and $\delta > 0$ sufficiently small and there will exist times τ_1 and τ_2 where $\tau_2 > \tau_1 + \delta$ such that

$$\tilde{R}_t^j > a \text{ on } [\tau_1, \tau_1 + \delta]$$

$$\tilde{R}_t^j > a \text{ on } [\tau_2, \tau_2 + \delta].$$

Now construct new paths which are identical to the old paths except on $[\tau_1, \tau_1 + \delta]$ where $R_t^i = \tilde{R}_t^i + a$ and $R_t^j = \tilde{R}_t^j - a$ and on $[\tau_2, \tau_2 + \delta]$ where $R_t^i = \tilde{R}_t^i - a$ and $R_t^j = \tilde{R}_t^j + a$. Then the change in $\mathscr{C}$ is easily calculated to be

$$\Delta\mathscr{C} = -\int_{\tau_1}^{\tau_1+\delta} (b^j - b^i) a e^{-rt} dt + \int_{\tau_2}^{\tau_2+\delta} (b^j - b^i) a e^{-rt} dt < 0.$$

Note that the depletion times of resource 1 and 2 are determined from the relationships

$$\int_0^{t_1} \bar{R}_t dt = \bar{X}^1 \tag{11}$$

$$\int_{t_1}^{t_2} \bar{R}_t dt = X^2 \tag{12}$$

Now we will characterize the shadow prices associated with the resource stocks by solving the problem formally. Nordhaus (6) has computed a set of energy resource shadow prices using linear programming and for a much more detailed model of costs and energy flows. The energy flows are exogenously specified and the present value of the future costs is minimized.

Form the Hamiltonian as

$$H_t = \left[\sum_{i=1}^{3} b^i R_t^i\right] e^{-rt} + \sum_{i=1}^{3} \mu^i R_t^i. \tag{13}$$

Necessary conditions which the efficient solution must satisfy are well known [e.g., see (1), Proposition 3, p. 38]. H_t must be minimized with respect to the choice of R_t^1, R_t^2 and R_t^3 where $\mu^i = 0$ if it is not efficient to deplete stock i and $\mu^i \geq 0$ if it is efficient to deplete stock i. In the latter case the $\mu^{i\prime}$s are chosen so that there is an incentive to use just the available amounts of the resources. Since H_t is linear in R_t^i, for H_t to be minimized it is necessary that

$$R_t^i = 0 \text{ if } c_t^i > c_t^{min} \tag{14}$$

where

$$c_t^i = b^i e^{-rt} + \mu^i \tag{15}$$

$$c_t^{min} = \underset{i}{\text{Min}} \{c_t^i\}.$$

The interpretation of these results is the following: c_t^i gives the present value cost associated with using an additional unit of R_t^i. The present value of the extraction cost is $b^i e^{-rt}$. The shadow price μ^i gives the present value of a marginal increase in resource stock i. Specifically,

$$\mu^i = -\frac{\partial \mathscr{C}}{\partial \bar{X}^i}.$$

Equation (14) states that only the least costly resources at time t will be used then.

The present value shadow prices μ^i will now be determined. We know that on the interval $[0, T]$, where t_1 is given by (11), only R_t^1 is used which implies that $c_t^1 \leq c_t^2$ and $c_t^1 \leq c_t^3$. Similarly on (t_1, t_2) it is necessary that $c_t^2 \leq c_t^1$ and $c_2^2 \leq c_t^3$, and on (t_2, T) it is necessary that $c_t^3 \leq c_t^1$ and $c_t^3 \leq c_t^2$. Since the $c_t^{i\prime}$s are continuous functions of time, at the depletion times

$$c_{t1}^1 = c_{t1}^2, \tag{16}$$

$$c_{t2}^2 = c_{t2}^3. \tag{17}$$

These imply

$$\mu^2 = (b^3 - b^2)e^{-rt2},$$

$$\mu^1 - \mu^2 = (b^2 - b^1)e^{-rt1} \tag{19}$$

from which μ^1 and μ^2 are determined and where $\mu^3 = 0$ since there is no value in increasing resource stock 3.

Hence, the difference in present value between resources 1 and 2 is equal to the extraction cost difference discounted from the depletion time of resource 1. Sometimes the $\mu^{i\prime}$s are called present value user costs. For example, μ_1 is the present value of the future cost incurred by using an additional unit of resource 1. Suppose resource 1 is depleted slightly sooner requiring one unit of resource 2 to be substituted at t_1. The present value of the additional cost is μ^2 plus the present value of the additional extraction cost which is $(b^2 - b^1)e^{-rt_1}$. The preceding analysis and interpretation can easily be extended to some finite number N of resource stocks.

Now consider a generalization for determining the resource efficiency price when extraction costs may increase continuously with time. A function b_t is well defined since it is efficient to use resources in the order of least extraction cost first. In the previous analysis b_t was required to be a step function. Now intervals will be admitted on which b_t is continuous and increasing.

The main problem is how to determine the present value shadow price associated with the resource of extraction cost b when b_t is continuous and increasing on an interval. Denote the shadow price assignment by $\mu(b)$. Recall that before the μ^i's were determined by (18) and (19) so that the following property would hold: If at time t it is efficient to use resource j, then

$$c_t^j = \underset{i=1,N}{\text{Min}}\,\{c_t^i\}. \tag{20}$$

Note that equation (14) implies that the minimization is taken over all resources including those fully extracted by time t. That is, it is not efficient to regret that a resource stock has been depleted.

Similarly, if the resource with extraction cost b is to be used at time t_b, then it is necessary that

$$c_{t_b}(b) = \underset{\beta}{\text{Min}}\,\{c_{t_b}(\beta)\}, \tag{21}$$

holding t_b fixed where

$$c_{t_b}(b) = be^{-rt_b} + \mu(b). \tag{22}$$

Therefore, if extraction cost is continuous and increasing on an interval containing t_b, a necessary condition for a local minimum is

$$\frac{\partial c_{t_b}}{\partial b} = 0$$

holding t_b fixed or that

$$\frac{d\mu}{db} = -e^{-rt_b}. \tag{23}$$

It is immediately seen that $\mu(b)$ is strictly decreasing and strictly convex on the interval since

$$\frac{d^2\mu}{db^2} = re^{-rt_b}\frac{dt_b}{db}$$

where t_b, the usage time for resource b, is determined as a function of b by the expression

$$\int_0^{t_b} \bar{R}_t \, dt = \Phi(b) \tag{24}$$

where $\Phi(b)$ gives the total initial resource endowment with extraction cost less than or equal to b.

Expression (23) may also be derived using a limit argument. Suppose b_t is continuous and increasing on an interval with range $(b, \hat{b})$. Partition the interval into N segments in such a way that the increments Δb are constant. The shadow prices of the N finite stocks are related by

$$\mu^j = \mu^{j+1} + (b^{j+1} - b^j)e^{-rt_j} \qquad j = 1, N \tag{25}$$

[see equation (19)] where t_j is the depletion time for resource stock j, $b^1 = b$ and $b^{N+1} = \hat{b}$. Solving equation (25) gives

$$\mu(b) = \mu(\hat{b}) + \sum_{j=1}^{N} e^{-rt_j} \Delta b.$$

Taking the limit as $N \to \infty$, the summation converges to

$$\mu(b) = \mu(\hat{b}) + \int_b^{\hat{b}} e^{-rt_\beta} d\beta, \tag{26}$$

where t_β is the time when resources of extraction cost β are used and is given by (24). Differentiating equation (26) implies that equation (23) must hold.

It will now be argued that the following holds even when b_t is not continuous and increasing:

$$\mu(b) = \mu(b_T) + \int_b^{b_T} e^{-rt_\beta} d\beta. \tag{27}$$

Suppose b_t is continuous, but constant on some intervals and increasing on others. Then the solution for $\mu(b)$ can be pieced together using result (26) on each interval on which b_t is increasing. Combining the results implies that equation (27) holds where in (27) t_β, still given by (24), is the depletion time for resource stock β. Now suppose there is a discontinuous jump from b^i to b^{i+1} at t_i. Then

$$\mu(b^i) = \mu(b^{i+1}) + (b^{i+1} - b^i)e^{-rt_i} \tag{28}$$

since at time t_i it is necessary that

$$c_{t_i}(b^i) = c_{t_i}(b^{i+1}).$$

If negligible amounts of resource were available with extraction cost b where $b^i < b < b^{i+1}$ its present value would be

$$\mu(b) = \mu(b^{i+1}) + (b^{i+1} - b)e^{-rt_i} \tag{29}$$

since the negligible amount does not affect the depletion time t_i. In this case $\mu(b)$ is a linear decreasing function of b with slope $-e^{-rt_i}$. Equation (27) assigns the value (29) to b, since on (b^i, b^{i+1}) t_β in (27) will be the constant t_i. Hence equation (27) completely characterizes the assignment of present value shadow prices to resources with extraction cost b for any nondecreasing piecewise continuous function b_t.

For illustration, suppose the resource flow is the constant $\bar{R}$. Then substituting t_b from equation (24) into (27) yields

$$\mu(b) = \int_b^{b_T} e^{-r\Phi(\beta)/\bar{R}} d\beta .$$

For the cost minimization problem with fixed planning horizon T and resource flow R_t exogenous, which is discussed in this section, it is necessary that $\mu(b_T) = 0$. However, when the resource extraction problem is embedded in a broader economic context, it may be that resources with extraction cost b_T do have economic value, in which case $\mu(b_T) > 0$. In the next section it will be shown that for an economic growth problem, present value resource shadow prices are assigned to resource cost grades according to equation (27) with $\mu(b_T) = 0$ where b_T will be equal to the unit cost of using a backstop technology. Of course, t_b in (24) will depend on the efficient resource flow path.

III. EFFICIENT ECONOMIC GROWTH

Solow and Wan (7) solve an efficient economic growth problem where resource extraction costs are increasing. They seek the largest level of consumption which can be sustained permanently given Cobb-Douglas technology with labor, capital, and resource factors. They prove that it is efficient to use the resources in order of increasing extraction cost. The efficient resource flow rate depends on a time varying shadow price. Heal (3) solves an optimal growth problem with the usual additive utilitarian welfare function with a positive discount rate on utility. He assumes a general aggregate production function F(K,R). He also assumes that resource extraction costs are increasing, but in contrast to Solow and Wan (7), he assumes a backstop technology is available which can produce a perfect substitute for the resource but without drawing on an exhaustible resource stock. Again the efficient resource flow depends on a time varying shadow price, which will be discussed further here. It appears in the paper that the welfare criterion plays a central role, disguising the efficiency nature of the resource pricing relationships. The following formulation is similar to Heal's.

Denote cumulative extraction from the resource stock by z_t so

$$z_t = \int_0^t R_\tau d\tau.$$

It will be assumed that it is efficient to use resources in the order of increasing extraction cost with no overlapping intervals. Then extraction cost is a nondecreasing function of cumulative extraction

$$b_t = g(z_t).$$

For now it will be assumed that $g(z_t)$ is continuous and differentiable. Removing this restriction will be discussed later. Note that $g(z)$ equals $\Phi^{-1}(z)$ given in equation (24) when the inverse exists. Let "a" be the cost of the backstop technology and define $\bar{z}$ so that

$$g(\bar{z}) = a. \tag{30}$$

Resources will be extracted until some time t_e at which $z_e = \bar{z}$ and then a switch will be made to the backstop technology. Let S_t be production from the backstop technology.

Assume that output is given by an aggregate production function $F(K, R + S)$ and output can be used for consumption, investment or resource extraction. Then the capital and cumulative extraction paths satisfy

$$\dot{K}_t = F(K_t, R_t + S_t) - C_t - g(z_t)R_t - aS_t \tag{31}$$

$$\dot{z}_t = R_t \tag{32}$$

with K_0 fixed and $z_0 = 0$.

Let C_t be a continuous, efficient consumption path defined on $[0, \infty)$. That is, C_t must be feasible and there must exist no other continuous, feasible path which is everywhere at least as good and which is better on some interval.

It will be assumed that eventually it will be efficient to switch to the backstop technology at some time t_e. Therefore, one can choose a planning horizon T sufficiently long so that $T > t_e$. It is necessary that R_t and S_t on $[0, T]$ be chosen to maximize K_T, given some continuous, efficient consumption path $\bar{C}_t$ on $[0, \infty)$. Otherwise, a consumption path $\bar{\bar{C}}_t$ could be constructed identical to $\bar{C}_t$ on $[0, T]$ but dominating $\bar{C}_t$ on $[T, \infty)$. Hence, it is necessary that the efficient flows R_t and S_t on $[0, T]$ maximize K_T subject to (31) and (32) with

$$z_T = \bar{z} \tag{33}$$

$$R_t \geq 0 \tag{34}$$

$$S_t \geq 0 \tag{35}$$

for all T sufficiently large.

The Hamiltonian for this problem is

$$H_t = \lambda_t[F(K_t, R_t + S_t) - \bar{C}_t - g(z_t)R_t - aS_t] - \mu_t R_t. \tag{36}$$

It is necessary that the efficient flows R_t and S_t maximize H_t where the shadow prices satisfy

$$\dot{\lambda}_t = -\frac{\partial H_t}{\partial K_t} = -\lambda_t F_K \tag{37}$$

$$\dot{\mu}_t = \frac{\partial H_t}{\partial z_t} = -\lambda_t g'(z_t)R_t = -\lambda_t g'(z_t)\dot{z}_t = -\lambda_t \frac{dg}{dt}. \tag{38}$$

Hence, if $R_t > 0$

$$\frac{\partial H_t}{\partial R_t} = \lambda_t[F_R - g(z_t)] - \mu_t = 0$$

or

$$F_R = g(z_t) + \mu_t/\lambda_t, \tag{39}$$

and if $S_t > 0$

$$\frac{\partial H_t}{\partial S_t} = \lambda_t[F_R - a] = 0$$

or

$$F_R = a. \tag{40}$$

First, the behavior of the model at the switch time t_e will be investigated. It is assumed that an instantaneous switch is efficient. Let $(R_e, 0)$ be the flow at te^- and $(0, S_e)$ be the flow at t_e^+. At t_e the Hamiltonian must be maximized with respect to (R_t, S_t). Then it is necessary that both $(R_e, 0)$ and $(0, S_e)$ yield maximum values for H. Equating these values (i.e., $H_e^- = H_e^+$) yields

$$\lambda_e[F(K_e, R_e) - \bar{C}_e - g(\bar{z})R_e] - \mu_e R_e = \lambda_e[F(K_e, S_e) - \bar{C}_e - aS_e]. \tag{41}$$

Substituting (39) into the LHS of (41) and substituting (40) into the RHS yields

$$F(K_e, R_e) - F_R(K_e, R_e)R_e = F(K_e, S_e) - F_R(K_e, S_e)S_e.$$

Since the function $F(K_e, x) - F_x(K_e, x)x$ is strictly increasing in x for $F_{RR} < 0$, clearly

$$R_e = S_e. \tag{42}$$

Then (41) implies

$$g(\bar{z}) + \mu_e/\lambda_e = a$$

Finally (30) implies

$$\mu_e = 0. \tag{43}$$

We now have the results needed for a careful look at the efficiency prices for the resources. First note that F_K is the return to capital and λ_t is the discount factor satisfying

$$-\frac{\lambda_t}{\lambda_t} = F\ . \tag{44}$$

If $R_t > 0$, from (39) the marginal product of the resource F_R must equal the extraction cost plus the shadow price of the resource at t given by μ_t/λ_t. Since λ_t is the discount factor, μ_t is the present value shadow price of the resource. It is claimed that the formula for assigning shadow prices to resource stocks with various extraction costs is the same in the growth context as the one developed in Section II. Integrating (38) yields

$$\mu(b_t) = \int_{b_t}^{a} -\lambda_{t_\beta}\, d\beta \tag{45}$$

where t_β satisfying (24) is the time when resources of extraction cost β are used and by (43) $\mu(a) = 0$. Of course t_β and hence $\mu(b_t)$ depend on the efficient resource path R_t. Equation (45) is identical to (27) except that the discount rate is time varying in (45). Hence, the μ_t may be given the same present value user cost interpretation discussed in Section II regarding the present value of the increased future extraction costs associated with using an additional unit of resource at t.

For $t < t_e$ the efficient resource flow R_t is determined from equation (39) in terms of the extraction cost and μ_t. Let the RHS of (39) be denoted by p_t, the efficiency price of the extracted resource. Then R_t is the solution to

$$p_t = F_R(K_t, R_t) \tag{46}$$

which takes the form of a demand function. Increasing K_t shifts the demand function to the right.

Finally, it will be shown that the rate of increase of the efficiency price for the extracted resource satisfies the same relationship as in the conventional theory with constant extraction costs:

$$p_t = g(z_t) + \mu_t/\lambda_t$$

$$\dot{p}_t = \frac{dg}{dt} + \frac{\dot{\mu}}{\lambda_t} + \left(-\frac{\dot{\lambda}_t}{\lambda_t}\right)\frac{\mu_t}{\lambda_t}$$

$$= \frac{dg}{dt} - \frac{dg}{dt} + F_K\frac{\mu_t}{\lambda_t}$$

$$\dot{p}_t = F_K(p_t - g(z_t)). \tag{47}$$

Hence, as was pointed out in the introduction, it cannot be efficient to have price stability and a significant resource value $p_t - g(z_t)$ simultaneously. However, recall that as z_t approaches $\bar{z}$ at t_e, μ_t approaches zero. Hence $\dot{p}_e = 0$, implying that the efficiency price of the extracted resource approaches the unit cost of the backstop technology in a stable manner. That is, for the time when all economically viable resources are depleted, the price is not only not rising fast, but it is not rising at all! For $t < t_e$, it can be shown that $p_t > g(z_t)$ and hence p_t is increasing. The current value of the unextracted resource $p_t - g(z_t)$ may be either increasing or decreasing, with the latter case associated with rapidly increasing extraction costs. Note that

$$\dot{p}_t - \frac{dg}{dt} = F_K(P_t - g(z_t)) - g'(z_t)R_t.$$

It can be shown that equations (45) and (47) still hold even if $g(z_t)$ is not differentiable or not continuous at a finite number of points. In the above problem $g(z_t)$ is not necessarily differentiable at t_e. Hence, two decision variables R,S are introduced representing production on $(0, t_e)$ and (t_e, ∞), respectively. The continuity of H at t_e links the solutions together. Specifically equations (42) and (43) hold. The same procedure may be used to handle the points where g(z) is not differentiable or continuous.

IV. COMPETITIVE MARKET

In this section it will be investigated whether the efficient pricing rule (47) is consistent with a market equilibrium. This question will be asked in a context even more simple than the growth model in Section III. Suppose that the resource demand function in known with certainty to be

$$R_t = D(p_t).$$

Further, suppose that there exists a minimum price p^c at which demand is reduced to zero. p^c is called a choke price.

Again it will be useful to think of extraction costs as a function of cumulative resource usage z_t with $b_t = g(z_t)$. It will be shown that the equilibrium involves using the resource with the least extraction cost first. Hence the function $g(z_t)$ is well defined. $g(z_t)$ is assumed to be known with certainty by economic agents.

It is claimed that the solution to the following problem is a competitive equilibrium with perfect foresight:

$$\dot{z}_t = D(p_t) \tag{48}$$

$$z_0 = 0 \tag{49}$$

$$g(z_e) \le p^c \tag{50}$$

$$p_t = r(p_t - g(z_t)) \tag{51}$$

$$p_e = p^c \tag{52}$$

with the condition that there are no remaining resources at t_e with extraction cost less than p^c. The solution consists of paths on some interval $(0, t_e)$ such that path p_t satisfies (51) with end point (52) and free p_0 and path z_t satisfies (48) with initial condition (49) and the largest end point z_e satisfying (50). Once the above conditions are met, t_e is determined. Note that for this t_e when all resources with extraction cost less than p^c are depleted, $p_e = p^c$ and demand is zero.

Suppose a firm owns a small amount of resource with extraction cost $\hat{b}$. The firm is a price-taker with a constant discount rate r. It is assumed that there are no common property externalities causing production to be too rapid. The question is, when should the firm sell its resource to maximize the present value of profits Π?

$$\Pi = (p_t - \hat{b})e^{-rt} \tag{53}$$

$$\begin{aligned} \frac{d\Pi}{dt} &= \dot{p}_t e^{-rt} - r(p_t - \hat{b})e^{-rt} \\ &= r(p_t - g(z_t))e^{-rt} - r(p_t - \hat{b})e^{-rt} \\ &= r(\hat{b} - g(z_t))e^{-rt}. \end{aligned} \tag{54}$$

It is argued that an equilibrium is to always use the remaining resource with the least extraction cost. In this case $g(z_t)$ is a nondecreasing function of time. Then by (54) profits are increasing until $g(z_t) = \hat{b}$ and decreasing thereafter. Hence, $\hat{b}$ will only be extracted if $g(z_t) = \hat{b}$ which is necessary to achieve the equilibrium that firms extract their resources in the order of increasing extraction cost.

Note that $g(z_t)$ may be constant on an interval. Then the conventional theory applies directly. Firms with resources of that extraction cost will be indifferent as to when they sell on that interval provided the price path satisfies equation (51). The resource flow $\dot{z}_t$ is determined by the demand function.

To summarize, firms with perfect foresight realizing the equilibrium is to extract the resources in the order of increasing extraction cost and at a rate just equal to demand, which depends on the price path (51), (52), do not have the incentive to deviate from the equilibrium behavior.

V. SUMMARY AND CONCLUSIONS

Suppose that resources with extraction cost $\bar{b}$ will eventually become marginally economical to extract and resources with $b > \bar{b}$ will not. Then clearly resources with $b < \bar{b}$ should command an economic value in excess of the cost of extraction. Suppose it is appropriate to discount future costs at a rate r. For convenience r is taken to be constant, but a time-varying rate adds nothing new.

Associated with using an additional unit of resource at t is not only its extraction cost, say b^1, but also a shadow price or user cost (previously denoted by $\mu(b^1)e^{rt}$). The user cost is the value at t of the increased costs of shifting the entire future extraction pattern slightly closer to t. That is, if cost grades are discrete, the switch to each higher cost grade is made slightly sooner. At each switch time an additional cost equal to the extraction cost difference is incurred. For example, at the first switch time t_1, when the resource with cost b^1 is depleted, the value at t of the additional extraction cost is $(b^2 - b^1)e^{-r(t_1 - t)}$. Then since resource b^2 is used slightly sooner, it is depleted slightly sooner at time t_2 and an additional cost $(b^2 - b^3)e^{-r(t_2 - t)}$ is incurred. If resources with $b \le \bar{b}$ are depleted over a finite period $[0, t_e]$ then the shifting stops at t_e with slightly less resource used at t_e. But since resources with cost $\bar{b}$ are economically marginal, there is no loss incurred. Hence, the value at t of the user cost of resources with cost b^1 is the summation over all the resource grades of the additional extraction cost discounted to t. When resources are available with a continuous range of extraction costs, summation is replaced with integration giving the result reported in section II. It is easy to see that the resulting user cost increases at a rate r. Hence, it is only necessary to compute the present value user cost $\mu(b^1)$ with the user cost at t equal to $\mu(b^1)e^{rt}$. This procedure can be used to generate the whole structure of user costs $\mu(b)$ for $0 < b < \bar{b}$. The $\mu(b)$ is a decreasing, convex function. Of course, the procedure depends on perfect knowledge of how much resource is available in each user cost grade and the usage rate, for these determine the depletion times of each cost grade and hence the appropriate discount factors used for computing user costs.

When the efficient level of resource flow R_t depends on total unit cost $p_t = b + \mu(b)e^{rt}$ where b is the least extraction cost among remaining resources, then R_t and $\mu(b)$ must be determined simultaneously. Solow and Wan (7) and Heal (3) have pointed out how to do this in a growth context. Heal's paper has been extended to include efficient but nonoptimal growth paths and a more general cost path b_t which need not be differentiable or continuous. An additional result is that the structure of $\mu(b)$ implies

$$\dot{p}_t = (p_t - b_t)$$

where b_t is the extraction cost of the resource currently used and $p_t - b_t$ is its current shadow price. This is the conventional market result when b_t is constant. However, it holds even if b_t is continuous and increasing. It is somewhat surprising that $\dot{p}_t$ does not depend on $\dot{b}_t$. Note that with this model, a large scarcity rent implies a rapidly increasing price. However, if b_t approaches $\bar{b}$ continuously, then the shadow price will approach zero and at this final point, the efficiency price of the extracted resource p_t will not be increasing.

An important question is to what extent the efficiency price of the extracted resource p_t tends to increase following the extraction cost path b_t and to what extent it deviates from it. A limited answer to this question can be given here. Integrating the above equation for $\dot{p}_t$ implies that the average resource shadow price must equal $(\bar{b} - p_0)/rt_e$ where t_e is the depletion time. If, for example, $rt_e = 5$, the average resource shadow price must not exceed 0.2 $\bar{b}$. However, if considerable time is required to accumulate production capacity for a substitute, then the resource shadow price may be significantly larger [see Hanson (2)].

Of course, a policy goal of keeping prices close to extraction costs may not be desirable in the real world, since the model analyzed here does not include several important features such as various sources of uncertainty, intergenerational equity and risk-bearing aspects, international market structure and some important microeconomic aspects of resource development efficiency.

REFERENCES

1. Arrow, K. J., and M. Kurz, *Public Investment, the Rate of Return, and Optimal Fiscal Policy*. The Johns Hopkins University Press, Baltimore, 1970.
2. Hanson D. A., "Competitive Price Behavior of an Exhaustible Resource Where the Rate of Substitution is Constrained," *International Economic Review*, 18 (1977), 135–149.
3. Heal, G., "The Relationship Between Price and Extraction Cost for a Resource with a Backstop Technology," *Bell Journal of Economics*, 7 (1976), 371–378.
4. Herfindahl, O. C., "Depletion and Economic Theory," in M. Gaffrey (ed.), *Extractive Resources and Taxation*. University of Wisconsin Press, Madison, 1967.
5. Hotelling, H., "The Economics of Exhaustible Resources," *Journal of Political Economy*, 39 (1931), 137–175.
6. Nordhaus, W. D., "The Allocation of Energy Resources," *Brookings Papers on Economic Activity*, 3, (1973), 529–576.
7. Solow, R. M., and F. Y. Wan, "Extraction Costs in the Theory of Exhaustible Resources," *Bell Journal of Economics*, 7 (1966), 359–370.
8. ———, "The Economics of Resources or the Resources of Economics," *American Economics Review*, 64 (May 1974), 1–14.

STAVING OFF THE BACKSTOP: DYNAMIC LIMIT-PRICING WITH A KINKED DEMAND CURVE

Stephen W. Salant, BOARD OF GOVERNORS OF THE FEDERAL RESERVE SYSTEM

I. INTRODUCTION

The successful manipulation of the world oil price by the OPEC cartel has stimulated a search among consuming nations for substitute energy sources and technologies. At the current real price of oil, few of these "backstops" can successfully compete. But should the oil price be raised sufficiently, other alternatives to oil might be profitably supplied. At prices where a new substitute could be profitably produced, demand for OPEC oil would be substantially more sensitive to price changes than in the current situation where little substitution is possible.

The question naturally arises as to how—in the presence of an existing,

Advances in the Economics of Energy and Resources, Vol. 2, pp. 187–204.
Copyright © 1979 by JAI Press, Inc.
All rights of reproduction in any form reserved.
ISBN 0-89232-079-6

high-cost, backstop technology—OPEC should manipulate the price over time to maximize the discounted stream of profits generated by the extraction and sale of its oil reserves. Should it, for example, sell all of its oil at prices strictly below the "limit price" at which the backstop would enter? Or should it sell for a stretch of time at the limit price? In the latter case, should it satisfy the entire market demand at that price or should it co-exist with the backstop technology?

The solution to such pricing problems is, of course, of interest to OPEC, and presumably the cartel's advisers have studied them in depth. But also, if one can justifiably assume that OPEC acts in an economically rational way, the solution to such problems will enable importing nations to forecast more accurately both the time when a backstop will take over and the pricing strategy which OPEC will follow in the interim.

Several years ago, Robert Solow (7, p. 3–5) opened the discussion by analyzing the effect of a high-cost backstop on the price of a low-cost exhaustible resource sold on a *competitive* market. Solow considered two cases. In the first, the backstop technology was assumed to be inexhaustible—his examples being solar energy and nuclear fusion. In the second case, the backstop itself was assumed exhaustible—his example being synthetic crude oil produced from ("liquefied") coal. In each case, Solow established that at first the low-cost, competitive extractors would supply the entire market while the price grew just enough to compensate them for not exchanging all of their underground stocks for interest-earning assets; once the limit price was reached, however, the backstop would enter and replace the low-cost extractors whose supplies would just, in equilibrium, be exhausted.

More recently Dasgupta and Stiglitz (2) and Stiglitz (8) have attempted to extend Solow's analysis to the case where the low-cost resource is owned by a *monopolist*. The backstop is assumed by them to be inexhaustible and the monopolist is assumed, in the tradition of von Stackelberg,[1] to take account of the supply responses of the backstop when formulating his optimal pricing strategy.

As I will show, there is a subtle but important error in the Dasgupta-Stiglitz analysis. Whether the backstop is inexhaustible or exhaustible, the optimal pricing strategy of the monopolist includes a phase where the monopolist charges the limit price but prevents entry by supplying the entire market himself. Neglect of this phase has resulted in *underestimates* of (a) the monopoly price which is currently optimal and (b) the time before the backstop will enter.

To simplify the exposition, the bulk of this paper concerns the case of an inexhaustible backstop. Section II introduces the monopolist's pricing problem and presents an intuitive solution to it. Section III verifies formally that the proposed solution is optimal. Section IV discusses the extension of the results to the case of an exhaustible backstop.

II. AN INTUITIVE ANALYSIS WHEN THE BACKSTOP IS INEXHAUSTIBLE

For simplicity, assume that an inexhaustible backstop can supply the entire market if the oil price reaches P_b but that it cannot compete at lower prices. By selling oil faster than the rate (Q^*) at which energy users wish to consume when charged the price P_b, the monopolistic extractor can create excess supply and depress the price below P_b. In equilibrium, the demand would then be exactly what it would have been with no backstop. By restricting sales to a lower rate than Q^*, however, the monopolist cannot raise the price above P_b as he could in the absence of a backstop; for the excess demand which would have developed will now be satisfied at the price P_b by the backstop suppliers. Figures 1 and 2 illustrate the situation. Figure 1 shows how the instantaneous demand and marginal revenue curves are altered by the presence of the backstop. Figure 2 shows how the associated total revenue function of the monopolist is affected.

If the original revenue function is continuous and strictly concave, the modified revenue function will be continuous and concave. Unlike the original function, however, the modified function will have a linear segment and a point where the derivative (marginal revenue) does not exist. The left-hand derivative at Q^* exceeds the right-hand derivative at that point. The total revenue lost by a unit reduction in sales below Q^* is equal to P_b while the total revenue gained by a unit increase in sales above Q^* is smaller than P_b. The asymmetry arises because the increase in sales, unlike the reduction, alters the price slightly and this alteration affects the revenue earned on the inframarginal units (Q^*).

The exact shape of the profit function depends on assumptions about

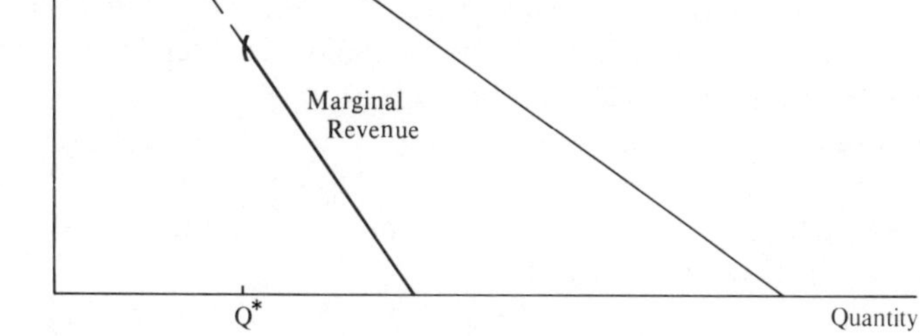

Figure 1.

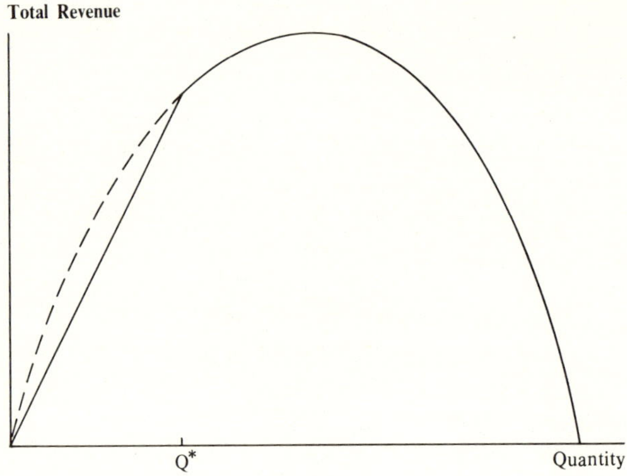

Figure 2.

the cost of extraction and the supply response of the inexhaustible backstop. If it is assumed that the monopolist has no costs of extraction, his profit function would be identical to the revenue function of Figure 2. Since the profit function would have a similar shape (a linear segment and a kink) if marginal costs were instead assumed to be a positive constant ($< P_b$), such a case would provide little to distinguish it from the case of zero costs. Hence we will not consider it further. If the monopolist's costs were assumed to be a smooth increasing, strictly convex function of the rate of extraction, his profit function would no longer contain a linear segment; instead it would be strictly concave. Even in such a case, however, the profit function would still have a kink. Indeed, such a kink would persist even if the supply response of the backstop were less extreme than we have assumed. Suppose, for example, that the backstop supply were zero for prices below P_b and increased gradually (rather than suddenly) at higher prices. Provided the backstop supply curve intersected the price axis at less than a vertical slope, the kink in the revenue function would still persist.[2] Since—in problems with a price-setting extractor and substitute technologies—nondifferentiable points[3] on the profit function appear unavoidable, their consequence for optimal extraction policies should be understood.

In the standard monopoly problem studied by Hotelling (4), the extractor begins with an initial inventory ($\bar{I}$), faces a stationary, strictly concave profit function ($\pi(Q)$), and picks a non-negative sales path ($Q(t)$) to maximize the integral of discounted profits:

$$\max_{\{Q\}} \int_0^\infty \pi(Q)e^{-rt}\,dt, \qquad \text{P1}$$

subject to constraints that:

$$Q \geq 0, S \geq 0,$$
$$\dot{S} = -Q,$$
$$S(0) = \bar{I}.$$

S(t) denotes the stock remaining underground at time t.

The profit function is assumed in the standard case to be differentiable everywhere. Provided the initial slope of the profit function ($\pi'(0)$) is finite, the optimal solution to this problem involves extracting the entire stock in finite time. Furthermore, for the program to be optimal, the discounted marginal profit must have a common value at each moment of positive extraction and no larger a value when zero extraction occurs. Any program violating these conditions can be dominated. For suppose, in a feasible program, that the discounted marginal profit from extracting positive amounts differed between any two moments.[4] Then the program under consideration could be dominated by an alternative feasible program where slightly more was sold at the moment with the higher discounted marginal profit and slightly less at the moment with the lower discounted marginal profit. Such an arbitrage could, of course, also be conducted if no extraction occurred at the moment with the higher discounted marginal profit.

In the presence of a backstop technology, the problem faced by the monopolist is similar to P1. The difference is that the instantaneous profit function has a kink at Q* and may have a linear segment to the left of Q*. At Q* the marginal profit is undefined. Denote the modified problem as P2.[5] Two cases might arise. If the profit function has a linear segment with a positive slope to the left of Q* but a negative slope for *all* extraction rates to the right of Q*, it is intuitively plausible that the monopolist should simply extract at the rate Q* for $\bar{I}/Q^*$ years and sell it at P_b.[6] If, however, marginal profit is positive for some extraction rates above Q*, it is less clear what extraction path is optimal.

It has been suggested[7] that the monopolist should set the value of marginal profit at a specified[8] initial level and should reduce extraction over time so as to maintain the same discounted value of marginal profit until the moment extraction falls to Q* and the price rises to P_b—at which point the monopolist should immediately bow out to the competition.

This proposed program is not optimal for P2, however, since it can be dominated. It involves a moment with positive extraction followed *immediately* by a moment of zero extraction but higher marginal profit. The lower marginal profit occurs the moment before extraction declines to Q* and the higher marginal profit occurs the moment after extraction falls below Q*.

If the proposed program is not optimal, what is?[9] No program where extraction exceeds Q* at one point and falls short of Q* moments later can be optimal since the marginal profit would jump up between the two moments, creating the opportunity for profitable arbitrage. The *optimal* program must avoid this situation by separating these two moments in time. If extraction is ever to exceed Q*, it should decline to Q*, *linger* there for an interval of time (θ) just long enough so that the left-hand derivative of profits at Q* ($\pi'(Q^*_-)$) discounted back θ periods is equal to the right-hand derivative at Q* ($\pi'(Q^*_+)$), and then decline below Q*.

The optimal length to linger at Q* can be computed from the parameters of the problem. In the absence of extraction costs, $\pi'(Q^*_-) = P_b$ while $\pi'(Q^*_+) = P_b\left\{1 - \frac{1}{|\eta(Q^*)|}\right\}$, where η is the price elasticity of consumer demand.[10] Hence, provided the extractor has sufficient reserves ($\bar{I} \geq \theta Q^*$), he should linger at P_b while supplying the entire market for θ years, where θ is defined by the following equation:

$$P_b e^{-r\theta} = P_b\left(1 - \frac{1}{|\eta(Q^*)|}\right)$$

$$\Rightarrow \theta = -\frac{1}{r}\ln\left(1 - \frac{1}{|\eta(Q^*)|}\right).$$

In the absence of sufficient reserves, the optimal strategy is to extract at the rate Q* for as long as possible and then shut down. The optimal length to linger is, therefore, min (θ, $\bar{I}/Q^*$).

To illustrate, suppose that the elasticity of demand for oil at the limit price were 2.0 and the real rate of interest were .05 per year. Then, provided the monopolist has sufficient reserves, he should gradually raise the price to P_b, supply the entire market at that price for approximately fourteen years, and then shut down. If a lower interest rate or elasticity of demand were assumed, the optimal length to linger would be longer.

Suppose, instead, that the cost of extraction were a positive smooth, increasing, strictly convex function of Q. In that case, the monopolist's profit function would still be kinked at Q* but it would instead be strictly concave. Since the marginal cost ($C'(Q^*)$) is continuous at Q*, both the left- and the right-hand derivative of the revenue function at that point would be reduced by the same amount. However, as a percentage of the right-hand derivative, the gap between the left- and right-hand derivatives at the kink would be larger than in the case without extraction costs. How would these modifications affect the optimal extraction policy? Once again, extraction should decline gradually to Q* so as to maintain the discounted value of marginal profit at some specified level; but then, extraction should linger at Q* *longer*[11] than in the previous case ($\hat{\theta} > \theta$) before resuming its decline. Furthermore, the abrupt decline from Q* to zero—which was optimal when the profit function had the linear segment—is no longer optimal since such a

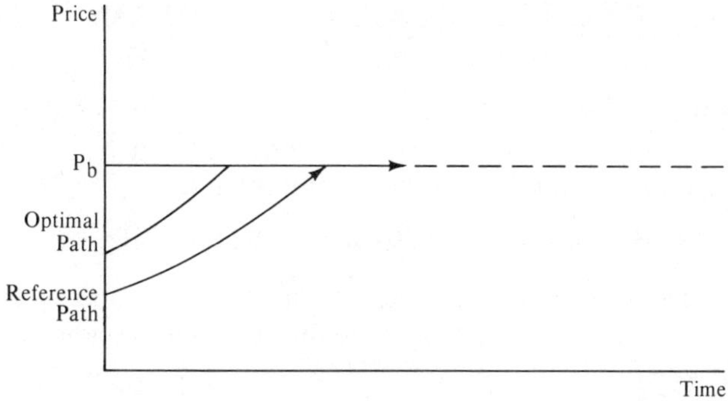

Figure 3.

reduction would result in a jump in marginal profit and the opportunity for profitable arbitrage. Instead, after lingering for $\hat{\theta}$ moments, the monopolist should begin again to reduce the rate of extraction so as to maintain the same level of discounted marginal profit until reserves are depleted. In this last segment of the program, the extractor and the backstop co-exist, each supplying part of the market.

In neither of the cases analyzed is it optimal to cease extraction the moment Q^* is reached. Since the monopolist needs some reserves for the phase which begins when Q^* is reached, less is available for the earlier phase of more rapid extraction. Consequently, the neglect of the final phase in each case has led to underestimates of the current profit-maximizing monopoly price. Furthermore, when the cost function is linear (constant or zero marginal costs), the neglect of the final phase has led to underestimates of the length of time before the backstop begins producing.[12] These points are illustrated in Figure 3. The arrowhead at the end of each path indicates the point when the monopolist exhausts his supplies and the backstop replaces him.

To understand why the backstop will not enter until later than was previously supposed optimal (see "reference path" of Figure 3), consider the following. Since the monopolist should set the initial price higher than was previously thought optimal, the subsequent price path should also be higher. The optimal price path will, therefore, reach the limit price and remain there for a while before the reference path catches up. At that time, it was thought that the monopolist would be out of reserves and that the backstop would, therefore, replace him. But since the monopolist will consistently have charged higher prices, less will have been demanded and he will in fact have some reserves left to extract. Hence, he can (and will) continue to stave off the backstop until his supplies are exhausted.

III. FORMAL VERIFICATION OF THE INTUITIVE ANALYSIS

To demonstrate formally that the strategies described above are indeed optimal, we verify that they satisfy conditions known to be sufficient for maximizing a functional. Define the Hamiltonian as

$$H = e^{-rt}\{\pi(Q) + \lambda(t)\dot{S}\} = e^{-rt}\{\pi(Q) - \lambda Q\},$$

where S is the state variable, λ the co-state variable, and Q the control. Any program optimal for P2 satisfies the following conditions for some continuous function $\lambda(t)$:

$$Q\ (\geq 0) \text{ maximizes the Hamiltonian at each instant (except on a set of measure zero),} \tag{1}$$

$$-\frac{\partial H}{\partial S} = \frac{\partial}{\partial t}\left[\lambda e^{-rt}\right] \Rightarrow \dot{\lambda} = r\lambda(\lambda \geq 0), \tag{2}$$

$$\dot{S} = -Q, \qquad (S \geq 0) \tag{3}$$

$$S(0) = \bar{I}. \tag{4}$$

Furthermore, it has been shown[13] that if the maximized Hamiltonian is concave in S, *any* program satisfying conditions (1) – (4) above and the following transversality condition must be optimal:

$$\lim_{t\to\infty} \lambda(t)e^{-rt} \geq 0 \text{ and } \lim_{t\to\infty} \lambda(t)e^{-rt}S(t) = 0. \tag{5}$$

Since, in our case, the maximized Hamiltonian is independent of S, it is (trivially) concave in S. Hence, to prove that a program is optimal, it is sufficient to establish that it satisfies conditions (1)–(5).

Even though the Hamiltonian is not differentiable everywhere with respect to Q, condition (1) can still be applied. That is, for each λ, the Q which maximizes the Hamiltonian can be determined.

We first consider the case with zero extraction costs.[14] In Figure 4 below, the functions $\pi(Q)$ and λQ—whose difference at each instant is proportional to the Hamiltonian—are graphed. The extraction rate which maximizes the Hamiltonian for a given λ can be expressed analytically.

Assuming that $\pi'(Q^*_+) > 0$,[15] we obtain:

$$Q(\lambda) = \begin{cases} X, \text{ where } \pi'(X) = \lambda, & \text{for } \pi'(Q^*_+) > \lambda \geq 0 \\ Q^* & \text{for } \pi'(Q^*_-) > \lambda \geq \pi'(Q^*_+) \\ Y, \text{ where } 0 \leq Y \leq Q^*, & \text{for } \lambda = \pi'(Q^*_-) \\ 0 & \text{for } \lambda > \pi'(Q^*_-). \end{cases} \tag{6}$$

Conditions (2) – (4) and (6) determine future values of λ and S for any initial assignment of λ. If we can find an assignment which also leads to

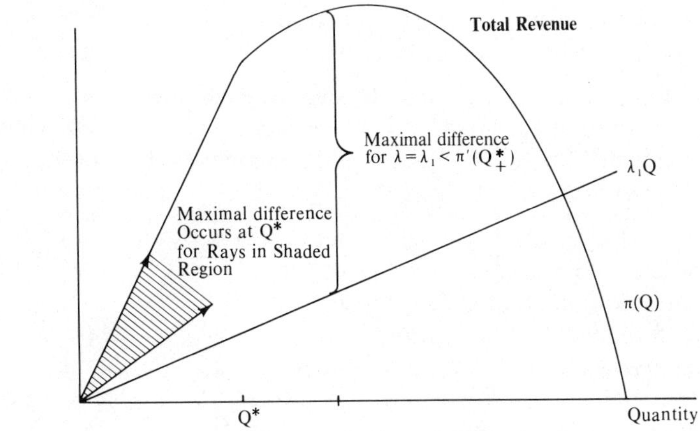

Figure 4.

satisfaction of the transversality condition, the associated program must be optimal. The search is facilitated by constructing the phase portrait sketched in Figure 5.

Equation (2) implies that there will be upward motion in the portrait above the horizontal axis but none on the axis. As for leftward motion, conditions (3) and (6) imply $\dot{S} = 0$ for the region where $\lambda > P_b (= \pi'(Q^*_-))$ and that $\dot{S} = -Q^*$ for the region where $P_b > \lambda \geq \pi'(Q^*_+)$. Since S can take on any of a set of values for $\lambda = P_b$, the horizontal motion is not well defined at the boundary of the two regions.

The initial value of the state variable is given by equation (4). Our only choice is in assigning an initial value to the co-state variable. What assignment will satisfy the transversality conditions (equation (5)) as well as the remaining conditions, which we have incorporated in the phase portrait?

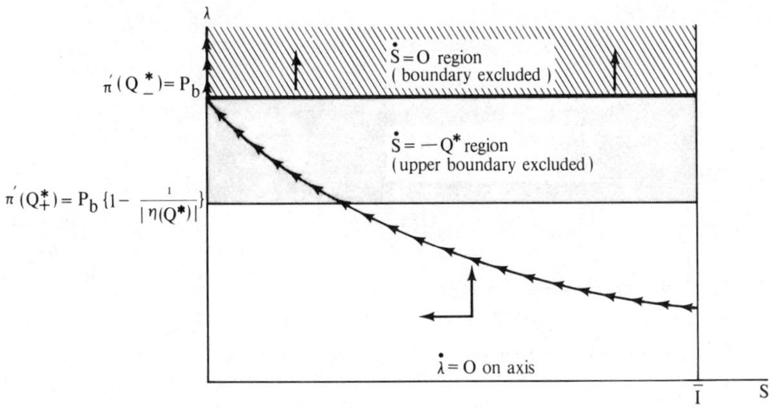

Figure 5.

Since $\dot{\lambda} = r\lambda$, $\lambda(t) = \lambda(0)e^{rt}$; therefore, the transversality condition can be rewritten as $\lim_{t\to\infty} \lambda(0)S(t) = \lambda(0) \lim_{t\to\infty} S(t) = 0$. If we initialize the co-state variable at zero, conditions (2) and (6) require that extraction continue foreover at a constant (positive) rate. However, such a program violates condition (3) which requires the stock remaining to be non-negative. Therefore, if the transversality condition is to be satisfied, $\lim_{t\to\infty} S(t) = 0$. However, for a trajectory to have this characteristic, it must pass into the $\dot{S} = 0$ region precisely when $S = 0$. Such a trajectory in fact exists and, since it satisfies each condition of the sufficiency theorem, it must be optimal.

We can determine the optimal trajectory by beginning at the point $(0, P_b)$ and working backward until the stock $\bar{I}$ is reached. If $\bar{I}$ is small, the value of the co-state variable when $S = \bar{I}$ may still exceed $\pi'(Q^*_+)$. This will be the case if $\bar{I} \leq \theta Q^*$. For larger initial stocks (the case illustrated in Figure 5), the initial value of the co-state variable for the trajectory passing through the point $(0, P_b)$ will be less than $\pi'(Q^*_+)$.

In either case, however, we have constructed an extraction program satisfying conditions (1)–(5). In the former case ($\lambda(0) > \pi'(Q^*_+)$), the optimal policy is to extract at the rate Q^* for $\bar{I}/Q^*$ moments, extract at any rate up to Q^* for the next instant, and then shut down. In the latter case ($\lambda(0) < \pi'(Q^*_+)$), the optimal policy is to extract at a gradually declining rate which initially exceeds Q^*; when Q^* is reached, extraction lingers at that rate for θ years, becomes momentarily indeterminate,[16] and then ceases altogether.

Few changes need be made to analyze the optimal strategy when cost is a smooth, increasing, strictly convex function of the rate of extraction. The strictly concave profit function ($\hat{\pi} = \pi - c$) replaces the concave revenue (π) function in equations (1)–(5). The rate of extraction ($\hat{Q}$) which maximizes the new Hamiltonian ($\hat{H}$) for given values of the co-state variable ($\hat{\lambda}$) has the following form:

$$\hat{Q}(\hat{\lambda}) = \begin{cases} X, \text{ where } \hat{\pi}'(X) = \hat{\lambda}, \text{ for } \hat{\pi}'(Q^*_+) > \hat{\lambda} \geq 0 \\ Q^*, \text{ for } \hat{\pi}(Q^*_-) > \hat{\lambda} \geq \hat{\pi}'(Q^*_+) \\ Y, \text{ where } \hat{\pi}'(Y) = \hat{\lambda}, \text{ for } \hat{\pi}'(0) > \hat{\lambda} \geq \hat{\pi}'(Q^*_-) \\ 0 \text{ for } \hat{\lambda} \geq \hat{\pi}'(0). \end{cases} \qquad (\hat{6})$$

Since the profit function is strictly concave and kinked at Q^*, $\hat{\pi}'(Q^*_+) < \hat{\pi}'(Q^*_-) < \hat{\pi}'(0)$. Depending on the nature of the cost function, various cases might arise. For example, if $P_b < C'(0)$ ($\hat{\pi}'(0) < 0$), the monopolist's optimal strategy would be never to extract. If, instead, $C'(0) < P_b < C'(Q^*)$, the monopolist should always extract less than the amount demanded at the back-stop price. The phase portrait of Figure 6 is drawn on the assumption that, in the (right-hand) neighborhood of Q^*, marginal extraction costs are smaller than marginal revenue ($\hat{\pi}'(Q^*_+) > 0$). In that case, a monopolist with

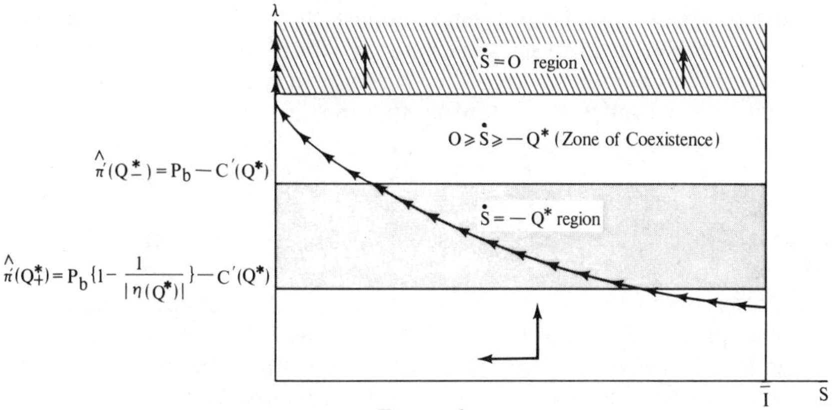

Figure 6.

an infinite stock to extract would find it optimal to produce at a rate greater than Q*.

In the case portrayed, the monopolist's initial stock ($\bar{I}$) is assumed to be large ($\bar{I} > \hat{\theta}Q^*$). The optimal trajectory passes through four regions. In the first, the monopolist raises the price over time until the backstop price is reached. In the second, he supplies the entire market at the backstop price for $\hat{\theta}$ periods. In the third, he gradually reduces his output and allows the backstop to take over. In the fourth, he shuts down. The coexistence phase occurs during the time when $\hat{\pi}'(0) > \hat{\lambda} > \hat{\pi}'(Q^*_-)$.

IV. EXTENSION TO THE CASE OF AN EXHAUSTIBLE BACKSTOP

If the resource supplied by the competitive sector is itself exhaustible, a complete characterization of the optimal path is more difficult. For, as Stiglitz (8, p. 659) has shown, the monopolist may then find it optimal to wait for the competitors to exhaust their supplies before beginning to extract his. We can hardly call the competitive sector a "backstop" in such a case; "forestop" might be a more appropriate label. To be a "backstop," the competitive sector must, in equilibrium, be the last to sell its resource. In the analysis below, we will restrict ourselves to those "backstop" equilibria where the backstop begins to extract only after the monopolist ceases to extract. In addition, we will assume that both resources are extracted at constant marginal costs. The reader can readily verify that the set of cases with these characteristics is not empty by considering the optimal strategy of a monopolist with zero extraction costs, facing a demand curve of constant elasticity. It can be shown[17] that if the competitive resource is costly to extract, the competitors will wait for the monopolist to exhaust his resource before beginning to extract theirs.

If the competitors find it optimal to go last, they will enter at a real price ($\bar{P}$) which enables them to sell their entire stock ($\bar{I}^c$) over time along a price path which rises enough to keep their discounted marginal profit constant. $\bar{P}$, therefore, must satisfy the following equation:

$$\bar{I}^c = \int_0^\infty Q^d\{(\bar{P} - k)e^{ry} + k\}\, dy,$$

where k is the constant marginal cost of the competitive extractors.

For the competitors to refrain from selling before $\bar{P}$ is reached, their discounted marginal profit must not then exceed its common value during the latter phase when they do sell. This limits the price the monopolist can charge in the earlier phase. For example, x moments before the entry of the competitors at $\bar{P}$, the monopolist cannot charge more than $(\bar{P} - k)e^{-rx} + k$. This ceiling can be viewed as a limit price for the case where the competitive backstop is exhaustible. In contrast to the inexhaustible case, this limit price rises over time. But as long as the marginal cost of the competitive extractors is positive, the limit price must rise by less than the rate of interest.

In what follows we will prove—for an equilibrium where the competitive extractors go last—that the monopolist will always find it optimal to linger for a while at this dynamic limit price while supplying the entire market. This constrasts with the proposition of Solow (7) that if the low-cost resource is competitively owned, the competitive extractors will shut down the moment the limit price is reached. The result of this section extends the conclusions of the previous sections by showing that limit-pricing remains the optimal strategy for a monopolist, even when the competitive backstop is exhaustible.

The demonstration utilizes Figure 7. In Figure 7, various paths terminating at $\bar{P}$ are drawn. Path I is the price path which would result if the monopolist equated discounted marginal revenue at every moment before $\bar{P}$ is reached. (In order for the proof to be nontrivial, path I is assumed to lie strictly below path IV to the left of their intersection at $\bar{P}$.) Path I is designated the "reference path."

To show that the optimal price path must coincide for at least some interval with the dynamic limit price, we show that all other price paths can be dominated. Suppose, for example, that the optimal price path were among the set of paths lying strictly below the dynamic limit price. This set contains not only the reference path but also paths along which the discounted marginal fluctuates. If the optimal path is anywhere in this set, it must be the reference path since all the other paths can—by a familiar argument—be dominated.

However, we will show that the reference path itself can be dominated. Consider the family of segmented price paths which result if the monopolist equates discounted marginal revenue at every moment until a particular

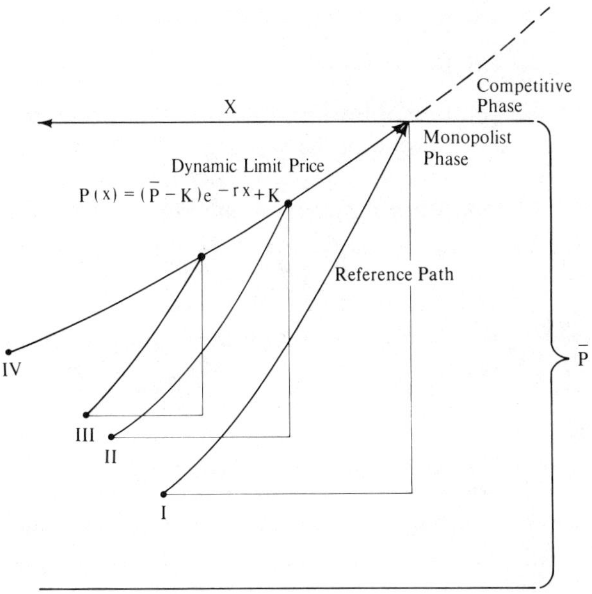

Figure 7.

point on the limit price path is reached, but which then coincide with the limit-price path. Each path is assumed to extend back from $\bar{P}$ far enough so that cumulative demand along its two segments exactly equals the monopolist's initial stock ($\bar{I}$).

This implies that the initial price on any upper path must exceed the initial price on any lower path (as is illustrated in Figure 7). For, suppose the contrary. Suppose some upper path extended to the left so far that its left-hand endpoint lay below the left-hand endpoint of some lower path. Then the upper path could be shifted laterally to the right so that the shifted path partially coincided with the lower path, extended beyond it to the left, and lay beneath it to the right before both paths reached $\bar{P}$. But it is then clear that cumulative demand along the shifted path exceeds cumulative demand along the lower path—contradicting the assumption on which Figure 7 is based.

Each member of the family of segmented price paths in Figure 7 can be characterized by a single number—the cumulative amount of reserves which the monopolist extracts after the limit price is reached. Path I can be seen as a limiting case where the cumulative amount extracted in the second segment of the price path is zero.

Denote the stock allocated to the second segment as S_2. For any allocation of total stock between the two segments, the monopolist's discounted revenue along the path may be written as:

$$V(S_2) = \{R_1(\bar{I} - S_2) + R_2(S_2)\}e^{-rT(S_2)}$$

where T (S_2) is the duration of the first segment of the program, $R_1(X)$ is the revenue (capitalized to the end of the first segment) of extracting X units of stock in the first segment, and $R_2(X)$ is the revenue (discounted to the beginning of the second segment) of extracting X units of stock in the second segment.

To prove that the reference path can be dominated, we simply verify that $V'(0) > 0$.[18] Since an optimal strategy exists but is not contained in the set of paths which lie strictly below the limit price to the left of $\bar{P}$, the optimal strategy must include a phase where the monopolist charges the (dynamic) limit price but prevents the backstop from entering by supplying the entire market himself.

Neglect of the lingering phase will once again result in *underestimates* of both the optimal current monopoly price and the date when the backstop will enter. Since the left-hand endpoint of each upper path must be higher than the left-hand endpoint of any lower path, the initial price along the

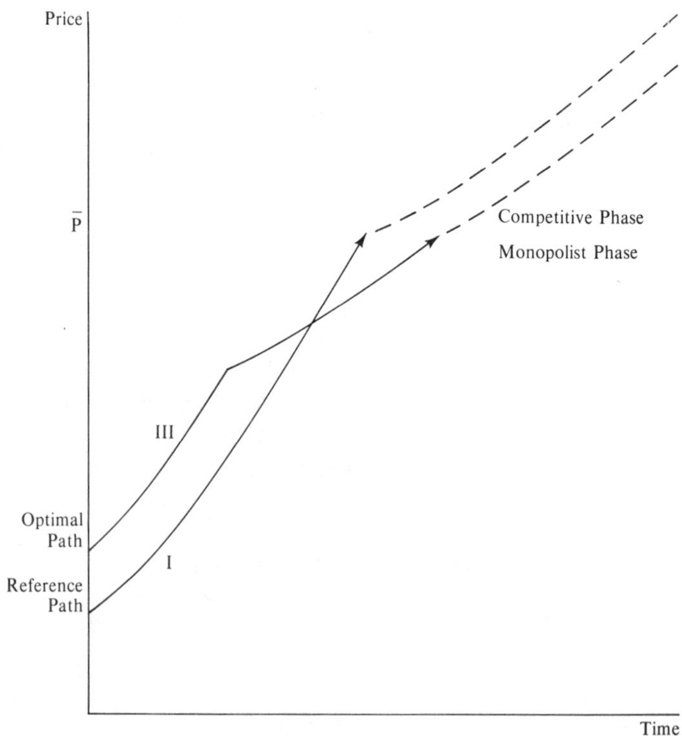

Figure 8.

optimal path must exceed the initial price along the reference path. Furthermore, since each path with a lingering phase lies above the reference path, the flow of demand will be less—x moments before $\bar{P}$ is reached—than it would be at the same point on the reference path. Therefore, in order for cumulative demand along these two paths to be equal, the optimal path must stretch back from $\bar{P}$ further than the reference path does. Hence, the backstop will enter later along the optimal path than it would along the reference path.

In Figure 8, the optimal path (assumed to be path III) and the reference path are plotted against time. Figure 8 is derived from Figure 7 by shifting path III to the right until its left-hand endpoint is aligned with the left-hand endpoint of the reference path. Figure 8, which is analogous to Figure 3, illustrates the two prediction biases which result if the reference path is mistakenly thought to be optimal.

In addition to the two biases discussed above, which have their counterpart if the backstop is inexhaustible, an additional bias will result when the backstop is exhaustible. Since the backstop enters later along the optimal path than it does along the reference path, the competitive rents accruing to owners of the backstop will be smaller than they would be if the reference path were optimal.

This paper has explored in detail a characteristic of cartel pricing problems which arises in the context of energy economics. But the characteristic—which may seem pathological because of its novelty—must arise whenever a leader takes account of a follower's optimal reaction and that reaction may lie on the boundary of a set restricted because of non-negativity conditions, capacity constraints, etc. In such cases, the leader's objective function has kinks and great care must be taken to determine what (if any) role such kinks play in the leader's optimal choice.

FOOTNOTES

I would like to thank Dale Henderson, Val Koromzay, and Janet Yellen for valuable comments on earlier drafts. This paper represents the views of the author and should not be interpreted as reflecting the views of the Federal Reserve System or other members of its staff.

1. An analysis of an exhaustible resource industry which utilizes the Cournot solution concept may be found in Salant (6).

2. That is, if

$$S(P) = \begin{cases} 0 \text{ for } P \leq P_b \\ f(P), \text{ for } P > P_b \end{cases}$$

(where f(P) is a smooth, monotonically increasing function with $f(P_b) = 0$ and $f'(P_b) > 0$), then the excess demand curve would have a kink at P_b. Its left-hand derivative would be $D'(P_b)$ while its right-hand derivative would be $D'(P_b) - f'(P_b)$. The associated revenue function would, therefore, also have a kink.

3. Gaskins (3) considered dynamic limit pricing when the monopolist controlled an inexaustible, but since he assumed the profit function of the monopolist had no kinks, the problem discussed here did not arise.

4. In continuous time, of course, *momentary* alterations in flow strategies have no consequences; each operation in the arbitrage must take place over a measurable interval of time.

5. In P2, $\pi(Q) = \min\{P_b Q, R(Q)\} - C(Q)$, where $R(Q)$ is the unconstrained, strictly concave revenue function and $C(Q)$ is a convex, extraction-cost function. It is assumed that a $Q^* > 0$ exists such that $\frac{R(Q^*)}{Q^*} = P_b$.

6. I am indebted to Heywood Fleisig for calling this case to my attention several years ago.

7. See, for example, Dasgupta and Stiglitz (2, p. 48). An allusion to the same analysis also appears in Stiglitz (8, p. 657).

8. The initial level of marginal profit is specified so that the initial stock is exhausted at the moment Q^* is reached.

9. In the problem at hand, an optimum exists if the discount rate is positive and the objective function is continuous. It is assumed that these conditions are satisfied.

10. The assumption that $R'(Q^*) > 0$ implies that $|\eta(Q^*)| > 1$. Hence the expression on the right-hand-side of the second equation is strictly positive.

11. If the monopolist has positive extraction costs and reserves sufficient for both the final coexistence phase and the full lingering phase, the optimal length to linger is $\hat{\theta}$, where $\theta = -\frac{1}{r}\ln\left(1 - \frac{P_b}{(P - C'(Q^*))|n(Q^*)|}\right)$. This formula can be deduced from the same considerations as were used when the marginal extraction cost was assumed to be zero. Indeed, the formula above reduces to the formula derived in the text ($\hat{\theta} = \theta$) if $C'(Q^*) = 0$. But if $C'(Q^*) > 0$, the argument in the function above is reduced. And since the negative log function is strictly decreasing, $\hat{\theta} > \theta$.

This result accords with intuition. If the marginal cost of extraction is positive, the marginal profit to the left and right of Q^* is reduced by the amount $C'(Q^*)$. Since the marginal profit to the right of Q^* is smaller, it takes a longer time ($\hat{\theta} > \theta$) to grow at the rate r by the same amount. It should be noted that the result ($\theta > \theta$) is a consequence of assuming a positive marginal cost of extraction and would hold even if the marginal cost were constant. Moreover, the result generalizes in the following way: the higher the marginal cost of extraction at Q^*, the longer the monopolist should linger there.

12. When the cost function is strictly convex, neglect of the lingering and coexistence phases will result in an underestimate of the time before the backstop supplies the *entire* market.

13. Arrow and Kurz (1, Proposition 8, p. 49).

14. As has been noted, the case with marginal cost a positive constant requires few changes.

15. If instead, $\pi(Q^*_+) \leq 0$ with π still concave, $Q(\lambda) = Q^*$ for $\pi'(Q^*_-) > \lambda \geq 0$, $0 \leq Q(\lambda) \leq Q^*$ for $\lambda = \pi'(Q^*_-)$, and $Q(\lambda), = 0$ for $\lambda > \pi'(Q^*_-)$. In this case, only the inelastic region of the demand curve is available to the monopolist and his best strategy is to sell at the rate Q^* until his supplies are exhausted.

16. The momentary "non-uniqueness of flow" in this problem has no consequence since trajectories which differ only for an instant will all result in the same integral of

discounted profits. Such an indeterminacy arises when the objective function has a linear segment, one familiar example being the so-called "bang-bang" solution of optimal control.

17. For, suppose the monopolist were the last to extract. If the monopolist could neglect the competitive response, he would enter at some point and supply the entire market while gradually reducing his extraction so as to raise marginal revenue at the rate of interest. Because the elasticity of demand is constant, however, this would cause the price to rise at the rate of interest. Foreseeing this rapid increase in prices, competitive extractors would calculate that by deferring extraction until later, they could increase their discounted profits *net* of extraction costs. Hence, their response to the proposed price path would be an unwillingness to extract anything before the phase of rapidly rising prices. But the monopolist is no fool and would foresee their response. In light of it, his optimal strategy—given that he goes last—is to sell along a price path where the competitors would just be willing to supply their resource first.

But even this best strategy when the monopolist goes last can be dominated by alternatives where the monopolist goes first. Consider the situation where the price path is identical to the one just described but now suppose the competitors extract *after* the monopolist. The competitors would be willing to delay extraction since they are just indifferent as to when they sell. However, since the monopolist has zero extraction costs, he gains from selling earlier.

In fact, the optimal strategy in this case is for the monopolist to sell first—along a path which rises at the rate of interest for a while but then *coincides* with a less steeply rising limit price. As we will show, the monopolist's optimal strategy will always involve continuing to sell after the limit price is reached.

18. Differentiating, we obtain:

$$V'(0) = \{R_2'(0) - R_1'(\bar{I})\}e^{-rT(0)} - rT'(0)\{R_1(\bar{I}) + R_2(0)\}e^{-rT(0)}$$

Selling a little less stock in the first segment and a little more in the second has two effects, which may be visualized by comparing path II to path I. As the first term in the expression above indicates, the reallocation raises the discounted profitability of the second segment and reduces that of the first, evaluated at the time when the two segments of the price path intersect. In addition, as is reflected in the second term, the reallocation of stock alters the duration of the first segment. To determine the sign of $V'(0)$, these various changes must be compared.

Since the left-hand endpoint of any upper path begins above the left-hand endpoint of any lower path, the upper path will reach the limit price sooner. This implies that $T'(S_2) < 0$. Since the revenue terms in the expression for $V'(0)$ are non-negative, $V'(0) > 0$ if $R_2'(0) > R'_1(\bar{I})$. The transfer of stock from the first to the second segment of the price path reduces R_1 and increases R_2. Since discounted marginal revenue is equated along the first segment of the price path, $R_1'(\bar{I}) = MR(Q^d(\bar{P}))$. Furthermore, the following argument establishes that $R_2'(0) = \bar{P}$.

By definition, the discounted profits of the monopolist during the second segment may be written as:

$$R_2(S_2) = \int_{x=0}^{\theta(S_2)} \{(\bar{P} - k)e^{-rx} + k\}Q^d\{(\bar{P} - k)e^{-rx} + k\}e^{r[\theta(S_2) - x]}dx,$$

where $\theta(S_2)$—the duration of the second segment—is implicity defined by

$$S_2 = \int_{x=0}^{\theta} Q^d\{\bar{P} - k)e^{-rx} + k\}dx.$$

From the second equation, it is evident that $\theta(0) = 0$. Differentiating the first equation in the neighborhood of $\theta = 0$, $R_2'(0) = \theta'(0)\bar{P}Q^d(\bar{P})$. Since $S_2'(\theta) = Q^d\{(\bar{P} - k)e^{-r\theta} + k\}$, $\theta'(0) = \frac{1}{Q^d\{\bar{P}\}}$. Hence, $R_2'(0) = \bar{P}$. Since $\bar{P} > MR(Q^d(\bar{P}))$, $V'(0) > 0$ as was to be shown.

REFERENCES

1. Arrow, Kenneth, and Kurz, Mordecai, *Public Investment, The Rate of Return and Optimal Fiscal Policy*. Johns Hopkins University Press, 1970.
2. Dasgupta, Partha, and Stiglitz, Joseph, "Uncertainty and the Rate of Extraction Under Alternative Institutional Arrangements," Technical Report No. 179 of the Institute for Mathematical Studies in the Social Sciences, March 1976.
3. Gaskins, Darius W., Jr., "Dynamic Limit Pricing: Optimal Pricing Under Threat of Entry," *Journal of Economic Theory*, 3 (1971).
4. Hotelling, Harold, "The Economics of Exhaustible Resources," *Journal of Political Economy* (April 1931).
5. Nordhaus, William D., "The Allocation of Energy Resources," Brookings Papers on Economic Activity, No. 3 (1973).
6. Salant, Stephen W., "Exhaustible Resources and Industrial Structure: A Nash-Cournot Approach to the World Oil Market," *Journal of Political Economy* (October 1976).
7. Solow, Robert M., "The Economics of Resources or the Resources of Economics," *American Economic Review* (May 1974).
8. Stiglitz, Joseph, "Monopoly and the Rate of Extraction of Exhaustible Resources," *American Economic Review* (September 1976).

ETA-MACRO: A MODEL OF ENERGY-ECONOMY INTERACTIONS

Alan S. Manne, STANFORD UNIVERSITY

I. ENERGY-ECONOMY INTERACTIONS

The ETA-MACRO model is designed to estimate the extent of two-way linkage between the energy sector and the balance of the U.S. economy. Implicitly, strong linkages are assumed both by proponents and by opponents of new energy supply technologies such as nuclear reactors and coal-based synthetic fuels. The proponents of these technologies argue that they are essential if the future rate of GNP growth is not to be slowed down. The opponents assert that the rate of growth *ought* to be slowed down—precisely in order to avoid the safety and environmental consequences associated with nuclear energy and with coal.

Advances in the Economics of Energy and Resources, Vol. 2, pp. 205–233.
Copyright © 1979 by JAI Press, Inc.
All rights of reproduction in any form reserved.
ISBN: 0-89232-079-6

Still another viewpoint is logically possible—one which rejects the hypothesis of a tight linkage between energy consumption and economic growth. According to this third viewpoint, there are many ways to reduce energy consumption without having a major impact on the GNP. In some cases, this would have no direct costs, but would require a change in life-styles, e.g., turning down thermostats to 65°F. during the winter. In other cases, it would be necessary to substitute other economic goods in place of energy. Three examples would be: (a) insulation to replace heating fuels in homes and other structures; (b) increased use of heat exchangers and of cogeneration within industry; and (c) the use of diesels to replace internal combustion engines in automobiles.

Each of these conservation measures would eventually become cost-effective if there were a sufficient rise in the price of energy. Opinions will differ on the magnitude of the price increase that would be needed in order to induce significant amounts of conservation. If one believes that cost-effective conservation is easy, the energy-GNP kinkage is a weak one. If, however, it turns out that the converse is true, the impending exhaustion of oil and gas resources leaves the United States with a difficult set of choices—either curtailing GNP growth or else actively promoting the introduction of nuclear energy and of coal-based technologies. Both solar and fusion are additional possibilities, but it seems unlikely that they can make a major contribution until the year 2020 or thereabouts.

Much of the energy-GNP linkage debate may be summarized in terms of a difference in views on the ease of substitution (in economists' language, the "elasticity of substitution") between energy and other inputs to the economy. In addition, one must allow for the possibility that a reduction in energy supplies or a rise in their price could lead to a drop in the rate of current output and investment—hence a reduction in the amount of physical capital that is available for use in future time periods.

ETA-MACRO is a single integrated model that incorporates the principal features that have just been noted: (a) the impending exhaustion of oil and gas resources—and a transition to new supply technologies over the next twenty-fifty years; (b) price-induced conservation—that is, the possibility of substituting other economic inputs in place of energy;[1] and (c) the effects of rising energy costs upon the accumulation of physical capital in future time periods.

Like most models, ETA-MACRO will lead to very different quantitative results, depending upon the specific numerical input assumptions. For the computer runs reported in this paper, the cost and supply technology assumptions were for the most part identical to those adopted by the CONAES Modeling Resource Group.[2] For short, these will hereafter be described as the MRG ground rules. *Other ground rules will lead to very different conclusions.*

A variety of energy sector policies may be explored in this way, but here we have focused upon just one specific policy option—banning the introduction of additional civilian nuclear power plants in the United States. Under base case assumptions, it turns out that a "no nuclear" policy would have negligible macroeconomic effects prior to the year 2000, but that it could lead to an *annual* loss of the order of \$100 billion by the year 2010. This is a large absolute number of dollars (approximately equal to today's entire U.S. military budget), but would still be a small percentage (about 3 percent) of the GNP in that year.[3]

In effect, ETA-MACRO permits us to estimate the magnitude of the insurance premium that would have to be paid in order to avoid the hazards (some real and others imaginary) of the civilian nuclear energy option. Is this insurance premium worth paying? Space does not permit us to address this topic here. Instead, we refer the reader to two reports (8, 14) with differing viewpoints on what may prove to be the key political issue—*not* reactor safety or waste disposal but rather the relation between civilian nuclear energy and weapons proliferation.

2. THE ECONOMY-WIDE AGGREGATE PRODUCTION FUNCTION

ETA-MACRO simulates a market economy through a dynamic nonlinear optimization process. Supplies and demands are equilibrated within each individual time period, but there are "look-ahead" features to allow for interactions between periods. For example, savings and investment decisions are modeled so that consumers will receive equal benefits from an additional dollar's worth of current consumption and a dollar's worth of investment. Before describing these relations in detail (the savings-consumption maximand), it seems best to begin with the two submodels of the production relationships within this economy: (a) ETA, a process analysis for energy technology assessment; and (b) a macroeconomic growth model providing for substitution between capital, labor and energy inputs.[4]

In order to focus upon the long-run issues of energy-economy interactions, resource exhaustion and the introduction of new technologies, the economy is described in highly aggregative terms. Outside the energy sector, all economic quantities are combined in terms of a *numeraire* (unit of account) based upon dollars of constant real purchasing power. Within the energy sector, only two categories are distinguished: electricity (without regard to the daily or annual load-duration curve) and nonelectric energy (combining liquids and gases on a BTU basis).

This classification system is too broad for many purposes. More details would be needed to analyze specific proposals for energy conservation or specific projects for the expansion of capacity within the electricity, petroleum

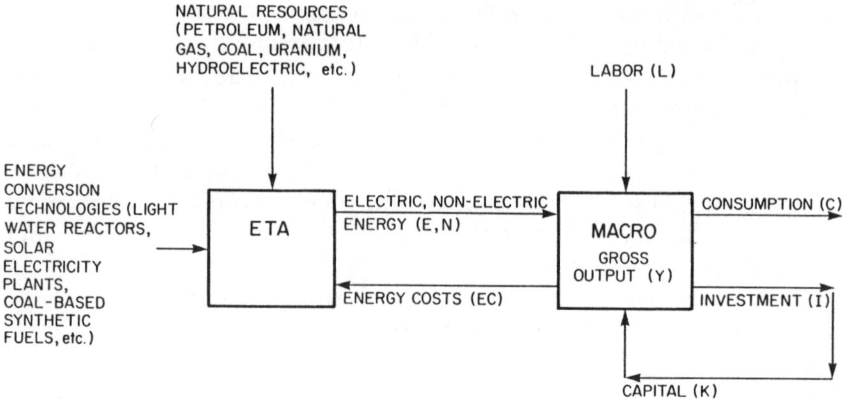

Figure 1. An overview of ETA-MACRO.

and gas industries. It is impractical, however, to construct a single model to answer all questions related to energy policy. Instead, it appears more reliable to depend upon *informal* information flows back and forth between individual analyses—each designed to handle specific issucs at an appropriate level of detail with respect to time, space, and physical product description.

Figure 1 provides an overview of the principal *static* linkages between the sectoral and the macro submodels. Electric and nonelectric energy (denoted, respectively, by the symbols E and N) are supplied by the energy sector to the rest of the economy. Like the material balance equations of an input-output model, aggregate economic output (Y) is allocated between interindustry payments for energy costs (EC) and "final demands" for current consumption (C) and investment (I). Thus:

$$\mathrm{Y} = \mathrm{C} + \mathrm{I} + \mathrm{EC}. \tag{1}$$

Each component of equation (1) is measured as an annual flow—in trillions of constant 1975 dollars. For the economy-wide production function, we assume that gross output (Y) depends upon four inputs: K, L, E, N—respectively, capital, labor (measured in "efficiency" units), electric and nonelectric energy. Moreover, we shall employ a nested nonlinear functional form based upon the following assumptions:

(a) There are constant returns to scale in terms of these four inputs;
(b) there is a *unit* elasticity of substitution between one pair of inputs—capital and labor—with α being the optimal value share of capital within this pair;
(c) there is a *unit* elasticity of substitution between the other pair of inputs—electric and nonelectric energy—with β being the optimal value share of electricity within this pair; and

(d) there is a *constant* elasticity of substitution between these two pairs of inputs—the constant being denoted by σ.

The long-run static production function may therefore be written:

$$Y = [a(K^{\alpha}L^{1-\alpha})^{\rho} + b(E^{\beta}N^{1-\beta})^{\rho}]^{1/\rho}, \tag{2}$$

where $\rho = (\sigma - 1)/\sigma \quad (\text{for } \sigma \neq 0, 1, \infty)$

The parameters a and b are two empirical constants that may be determined through a first-order optimality condition[5] for a year in which each of the four inputs (K, L, E and N) has been optimally adjusted to the prevailing prices. This equilibrium year is taken to be 1970. Clearly, the U.S. economy had *not* adjusted by 1975 to the energy price increases of 1972–1974.

To allow for time lags in the economy's response to higher energy prices, we can no longer work with a static model but must now introduce a time index t. A seventy-five-year planning horizon is employed, starting with 1975 and extending through 2050. There are sixteen individual time periods, each five years in length, and each centered around a "representative" year: 1975, 1980, ... 2050. With 1975 as year 0, therefore the time index t = 0, 5, ... 75.

To distinguish between short- and long-run responses to higher energy prices, we employ what is sometimes termed a "putty-clay" model. It is assumed that there is no flexibility with respect to the method of operation of the initial (1970) energy-using stocks and life-style habits, but that there is full flexibility with respect to capital goods accumulated after 1970. This probably overstates the rigidities with respect to the initial capital stocks (hard-baked "clay"), and it overstates the flexibility with respect to subsequent additions (the more malleable "putty").

With this type of model, the speed of adjustment depends upon a parameter to which we shall refer as SPDA. This constant appears in the capital accumulation equations below, and it defines the fraction of the initial capital stock that survives after one year of use. In general, we have supposed that capital stocks are replaced at the rate of 4 percent per year, and that the annual survival factor SPDA is therefore .96. This means that only 29 percent of the 1970 capital stocks will survive to the year 2000, and that all other capital goods will be appropriately adapted to the post-1970 increases in energy prices.

The *surviving* quantities are the following known values:

$$YS(t) = Y(1970)(SPDA)^{t+5} = 1.4146\ (.96)^{t+5} \quad (\text{unit: } \$10^{12}/\text{year})$$
$$KS(t) = K(1970)(SPDA)^{t+5} = 3.4\ (.96)^{t+5} \quad (\text{unit: } \$10^{12})$$
$$LS(t) = L(1970)(SPDA)^{t+5} = 1.0\ (.96)^{t+5} \quad (\text{unit: ratio to 1970, in ``efficiency'' units})$$
$$ES(t) = E(1970)(SPDA)^{t+5} = 1.65\ (.96)^{t+5} \quad (\text{unit: } 10^{12}\ \text{KWH/year})$$
$$NS(t) = N(1970)(SPDA)^{t+5} = .455\ (.96)^{t+5} \quad (\text{unit: } 10^{17}\ \text{BTU/year})$$

Let the unknowns YN(t), KN(t), ... denote, respectively, the gross output, capital stocks, etc., associated with the *new* investments made since 1970. By definition:

$$\begin{aligned} YN(t) &= Y(t) - YS(t) \\ KN(t) &= K(t) - KS(t) \\ LN(t) &= L(t) - LS(t) \\ EN(t) &= E(t) - ES(t) \\ NN(t) &= N(t) - NS(t). \end{aligned}$$

Total output may be calculated by adding two components—the first with fixed coefficients for the inputs of capital, labor and energy, and the second with flexible coefficients. From equation (1), we then have:

$$\begin{aligned} Y(t) &= YS(t) + YN(t) = \text{gross output at time t} \qquad (3) \\ &= YS(t) + [aKN(t)^{\rho\alpha} LN(t)^{\rho(1-\alpha)} \\ &\qquad + bEN(t)^{\rho\beta} NN(t)^{\rho(1-\beta)}]^{1/\rho}. \end{aligned}$$

In the production function (3), note that Harrod-neutral labor productivity growth is introduced through describing labor inputs L(t) in terms of "efficiency" units relative to a 1970 base. This means that *if* energy prices had remained constant from 1970 onward and *if* capital and energy inputs had increased at the same rate as the labor force index, then gross output would also have increased at this rate. *The labor force growth index is a key parameter*, and has a major influence upon the economy-wide potential growth rate. With higher energy prices, this full potential cannot be achieved, and gross output does not increase as rapidly as the labor force index. Following the CONAES Modeling Resource Group's base case GNP assumptions—but rounding and smoothing—all of the computer runs reported in this paper are based upon the following rates for the labor force and productivity growth index:

1970–1975 : 2.0% annual growth
1975–1980 : 4.5%
1980–1990 : 3.5%
1990–2000 : 3.0%
2000–2010 : 2.5%
2010–2050 : 2.0%

The following values have been adopted for the remaining parameters in the production functions (2) and (3):

$$\begin{aligned} \alpha &= .333 = \text{capital's value share} \\ \beta &= .40 \ \ = \text{electricity's value share} \\ \sigma &= .25 \ \ = \text{elasticity of substitution between} \\ &\qquad\quad \text{energy and non-energy inputs}^{6} \\ \therefore\ \rho &= (\sigma - 1)/\sigma = -3.0. \end{aligned}$$

As a general check for reasonableness, one may calculate the marginal productivity of capital in the base year 1970. From equation (2), the gross return on capital (before corporate income taxes and depreciation charges) is:

$$\frac{\partial Y}{\partial K} = Y^{1-\rho} K^{\alpha\rho - 1} a\alpha = 13.3\%/\text{year}.$$

3. ESTIMATING THE ELASTICITY OF SUBSTITUTION

Our best estimate is that σ (the elasticity of substitution) is of the order of .25. What is the empirical basis for this estimate? To a local approximation, it can be shown that σ is identical to the absolute value of the price elasticity of demand for primary energy. [See Hogan and Manne (6).] Unfortunately, there is a wide margin for error in the econometric estimation of this elasticity parameter, whether the statistical sample be time series or international cross-country comparisons.

Consider, for example, the evidence from the same period that was employed by Hudson and Jorgenson (7) for the estimation of price-responsive input-output coefficients. Figure 2 provides a scatter diagram of the time series for the United States during the years after World War II, 1947–1975. These years permit us to avoid confounding the issue of interfuel substitution (between coal and petroleum) with that of growth in the aggregate demand for energy. Prior to World War II, over 50 percent of total primary energy was derived from coal. It was not until the postwar period that oil and gas had replaced coal in most of its earlier uses, e.g., space heating and steam traction for railroad locomotives.

In calculating energy consumption, it seems appropriate to add oil and gas consumption together on a BTU basis. Surely this is inappropriate, however, for coal. As boiler fuels in electric utilities, oil and coal are virtually perfect substitutes. Nonetheless, the price that utilities have been willing to pay per BTU of coal has been consistently less than that for oil. Accordingly, for purposes of Figure 2, the aggregate consumption of fossil energy has been defined as the sum of oil and gas (measured in quads) plus 70 percent of the quads of coal consumed in the United States. This is sometimes termed a "value-weighted" measure of energy consumption. [See Mitchell (11).]

In order to isolate price from income effects, fossil energy consumption is expressed as a ratio to the GNP (in 1975 dollars) for each of the years shown in Figure 2. Note that the energy-GNP ratio has varied within relatively narrow limits—despite a 50 percent variation in the price of energy (expressed in dollars of constant purchasing power) during this period.

Figure 2 does *not* mean that there is an immutable energy-GNP ratio. Common sense examples (insulation, cogeneration, etc.) would suggest that

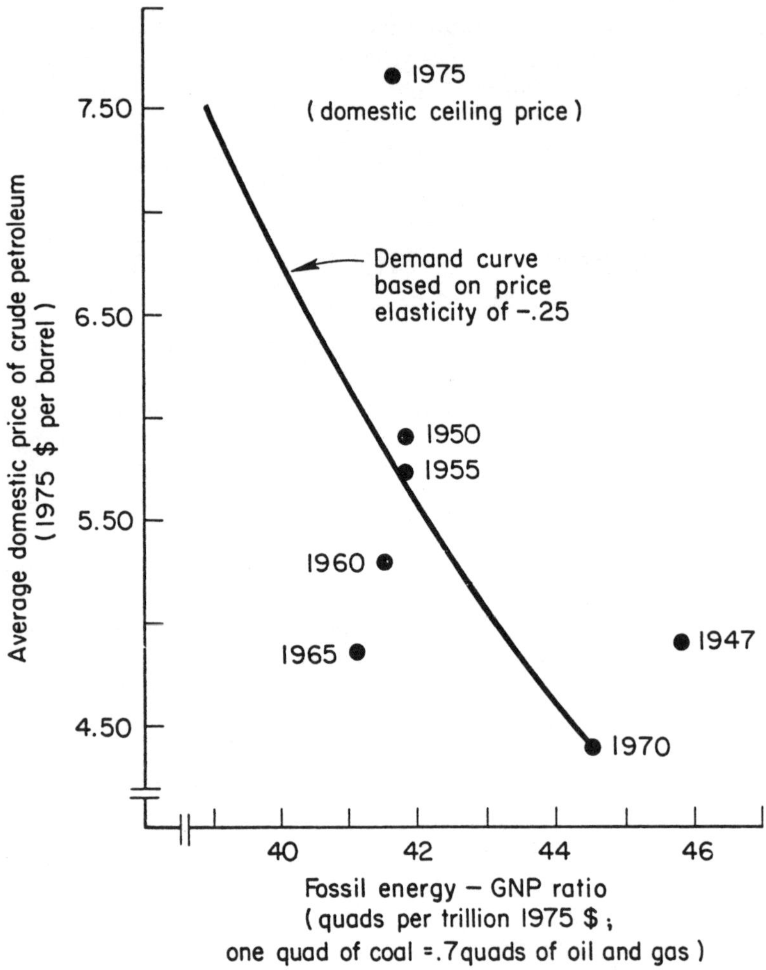

Figure 2. Price-consumption scatter diagram for fossil fuels, U.S., 1947–1975.

the price of energy could affect this ratio. What Figure 2 does suggest is that we ought to be quite cautious in our claims of reliability for one or another estimate of price elasticities. These econometric uncertainties would appear to be at least as great as the geological uncertainties on petroleum and uranium resources.

For reference purposes, one possible price-quantity demand curve is shown on the scatter diagram of Figure 2. It is drawn up on the same basis as the long-run production function, equation (3). That is, it represents an extrapolation based upon the actual 1970 energy price and energy-GNP ratio—assuming a long-run price elasticity of −.25. No claims of high reliability are made, but the time series evidence suggests that this does not represent a serious *underestimate* of the long-term price elasticity for primary energy within the United States.

One concluding comment on the measure of domestic energy prices. No one statistical concept is altogether appropriate. Coal and natural gas are particularly difficult because of their high transport costs—hence wide geographical price differences from one region of the country to another. For this reason, crude oil wellhead prices were taken as the most representative indicator of the price of primary energy in the United States during the postwar period.[7]

4. MACROECONOMIC RELATIONS BETWEEN INVESTMENT AND CONSUMPTION

In addition to equations (1) and (3), the macro submodel contains equations for physical capital accumulation and for terminal investment requirements. To approximate a two-year average gestation lag between investment and usable capital stocks, it is supposed that 60 percent of gross investment provides an immediate increase in the capital stock, but that 40 percent has a five-year delay. Capital stocks are expanded by gross investment I(t), and are reduced by the SPDA survival fraction. The capital accumulation equations are then:

$$K(t) = (SPDA)^5 K(t-5) + 2I(t-5) + 3I(t) \quad (t = 5, \ldots 75). \tag{4}$$

The 1975 initial conditions (in $\$10^{12}$) are:

$$\begin{aligned} I(0) &= .2 \\ C(0) &= 1.3 \\ K(0) &= 2.5[I(0) + C(0)] = 3.75. \end{aligned} \tag{5}$$

Gross investment during the terminal year (2050 = year 75) must provide for subsequent expansion at the annual rate g, and must also provide for replacement at the rate (1 − SPDA). Therefore:

$$I(75) = (g + 1 - SPDA)K(75). \tag{6}$$

For the post-terminal growth rate g, we have taken 2 percent per year—that is, the same rate as the expansion of the labor force (measured in efficiency units) from 2010 onward. This constraint reduces, but does *not* eliminate "horizon" effects. For this reason no numerical results are presented for the years after 2030.

Appendices A and B provide complete documentation on the remainder of ETA-MACRO. Each constraint in Appendix A is *linear* in terms of the unknowns. Equations (4)–(6) are also linear, and are included in the constraint set of ETA-MACRO.

There is a minor technical difficulty in the case of equations (1) and (3). These constraints include nonlinear terms and cannot be handled directly by the Murtagh-Saunders reduced gradient optimization program MINOS.[8] We may, however, eliminate the unknowns Y(t) by substituting equations (3) into (1), but noting that Y(t) is in turn a *non*linear function of the unknowns K(t), E(t) and N(t).[9] Aggregate consumption in period t is then:

$$C(t) = Y(t) - I(t) - EC(t). \tag{7}$$

For optimizing the pattern of investment and consumption over successive time periods, we shall take a special case of a criterion termed the discounted "utility" of consumption. The one-period utility discount rate[10] will be denoted by δ, and the utility function will be the logarithm of consumption—perhaps the simplest and the oldest member of the isoelastic class.[11] The maximand is then written:

$$\sum_{t=0.5}^{70}\left(\frac{1}{1+\delta}\right)^{t}\log C(t) + \sum_{t=75,80}^{\infty}\left(\frac{1}{1+\delta}\right)^{t}\log\left[C(75)(1+g)^{t-75}\right]. \tag{8}$$

Note that the rightmost sum in the maximand (8) includes the discounted utility of consumption during the post-terminal period – that is, from year 75 onward—assuming that the economy is capable of growth at the annual rate of g thereafter. Note also that this rightmost sum is a linear function of log C(75). For our base case calculations, we shall adopt a value of $\delta = 10$ percent per year, but will also show the implications of two other rates: 8 percent and 12 percent. The lower the utility discount rate, the more weight is placed upon distant future consumption goals, and the higher becomes the near-term optimal rate of investment. Thus, the δ parameter is closely related to the optimal savings rate in this economy. No other constraints are imposed upon the propensities to save and to consume. This means that the macro model is driven by the following three key inputs:

(a) the savings-investment accumulation process—primarily determined by δ;
(b) the labor force growth index L(t); and
(c) σ, the elasticity of substitution—the principal factor governing the economy's ability to cope with higher energy prices.

5. THE ETA SUBMODEL—PRINCIPAL ASSUMPTIONS

In its broad outlines, the ETA submodel follows the original version in Manne (10). The numerical parameter values here, however, have been those adopted by the CONAES Modeling Resource Group. These parameters include, for example, capital cost factors, fuel costs, dates and rates of introduction of new technologies, and the supply curves relating cumulative extraction to the minimum selling price of exhaustible resources such as petroleum and uranium.

Table 1 includes a complete list of all technologies considered here. The following departures have been made from the "base case" defined by the MRG:

(a) ETA-MACRO *includes* the MRG's limits on both coal (50 quads)[12] and shale oil (12 quads);
(b) petroleum resources now include two cost categories—respectively, \$2 and \$3 per million BTU; altogether, a total of 2,720 quads of domestic and imported petroleum and natural gas are available at a \$3 selling price;
(c) the natural uranium resource supply curve is *not* extended beyond the cumulative total of 3.7 million short tons of U_3O_8 available at \$45 per pound;

Table 1. Identification of technologies.

Technology number	*Name j*	*Identification*
Electric		
01	HYDR	hydroelectric, geothermal, wind, etc.*
02	RFOS	remaining initial fossil electric*
03	COLL	coal-fired electric, low-cost regions
04	COLH	coal-fired electric, high-cost regions
05	LWRA	LWR, no plutonium recycle, .2% tails
06	LWRB	same as LWRA, but applies to cases involving no new nuclear reactors
07	LWRC	LWR, plutonium recycle, .2% tails
08	FBRX	liquid metal fast breeder reactor, early oxide
09	ADVX	advanced solar electric or fusion
Nonelectric		
10	PETG	petroleum and natural gas
11	SYNF	coal-based synfuels (liquids and gases)
12	SHAL	shale oil
13	ELHY	electrolytic hydrogen
14	NAE5	nonelectric alternative energy systems—equivalent to oil and gas at \$5/million BTU
15	NAE8	same as NAE5, but applies to cases involving costs of \$8/million BTU
16	CLDU	coal—direct uses other than electricity and synfuels*

*Exogenously determined activity levels.

(d) the advanced converter (HTGR with 82 percent conversion ratio) is not an available option; and
(e) the MRG's 10 percent nonprice conservation factor is dropped.

The principal effect of these changes is that electricity demands are generally higher[13], and that the fast breeder reactor (FBRX) is a more competitive technology than was assumed by the MRG. Three of these changes (a, b and e) also mean that a "no nuclear" policy would have a greater impact on the economy than under the MRG's base case assumptions.

The *cost of capital* poses a difficult issue in any economic model. Here we have followed the MRG by taking 13 percent per year as the real cost to electric utilities—net of inflation and before corporate income taxes. This may have been representative of capital costs during the 1974–1975 recession, but appears to be on the high side for other time periods in the United States.

The 13 percent rate enters into ETA-MACRO through the levelized capital charges for each of the electricity production technologies. (See the cost coefficients cur(j) that appear in the energy cost equations COSTEN(t) in Appendix A.) In effect, the sector is viewed as *renting* capital goods from the rest of the economy and paying an annual return of 13 percent for these services.

Note that the corporate cost of capital is an altogether different concept from the "utility" discount rate δ that appears in the maximand, equation (8). The principal criterion for assigning a numerical value to δ is so that—in conjunction with the logarithmic utility function—it will lead to a plausible macroeconomic pattern of savings and investment.

Both δ and the corporate cost of capital are in turn different from a third concept that will be introduced below—the discount rate to be employed for *public* decisions. *If* there were no corporate income taxes or capital market imperfections, the public discount rate would always be at least as great as δ in a growing economy, i.e., one with a declining marginal utility of consumption.

In practice, however, there are wedges between the rates at which corporations are willing to invest, and the rates at which consumers are willing to save. In a second-best world, the public discount rate could therefore be considerably lower than δ.

6. SOME KEY NUMERICAL RESULTS

Before presenting detailed numerical results, it is worth reviewing the MRG ground rules. According to these rules, nuclear energy enjoys no significant cost advantage over coal-fired plants for base-load electricity generation. Thus, for as long as coal supplies are adequate to handle the electricity load growth, there can be no economic loss associated with a "no nuclear"

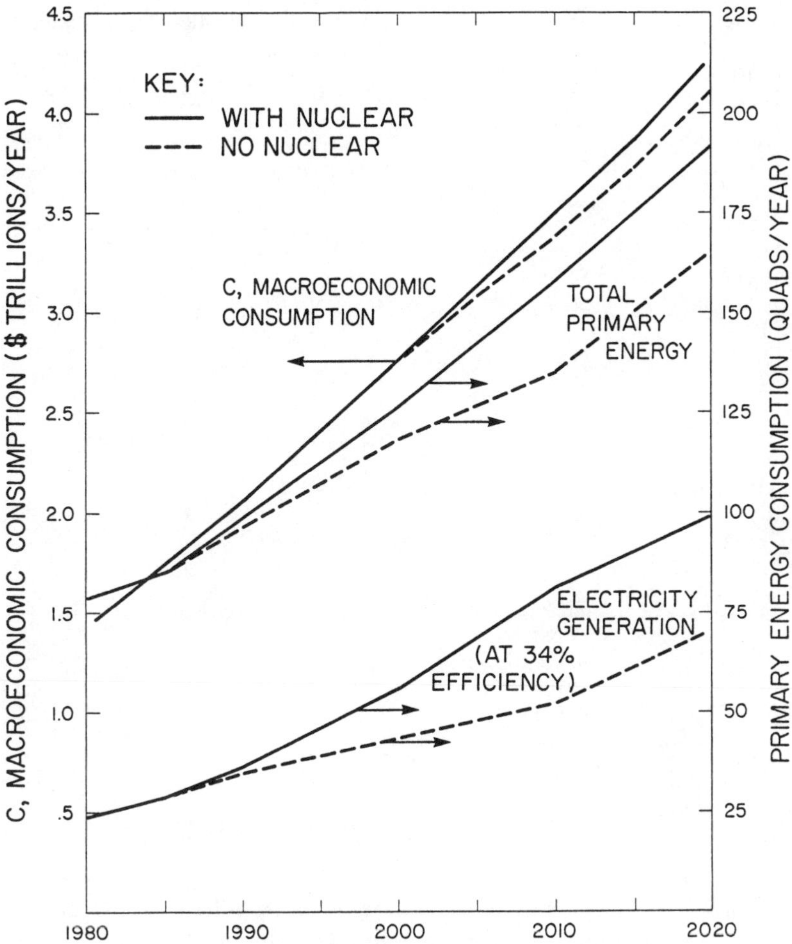

Figure 3. Base case results.

energy policy.[14] Under base case assumptions, it turns out that coal supplies would *not* be a binding constraint until after the year 2000, and that therefore aggregate consumption is unaffected by a "no nuclear" policy until that date. (See Figure 3.)

After the year 2000, the MRG ground rules provide that solar electricity is available as a "backstop" technology—but with levelized costs that are 20 mills/kwh higher than coal or nuclear. In the year 2010, for example, direct solar electricity would generate .65 trillion kwh under a "no nuclear" scenario. Thus, the direct costs of employing this backstop are $13 billion in that year. Adding together the other direct and indirect losses throughout the economic system via ETA-MACRO, it is not surprising therefore that

Table 2. Macroeconomic results – three alternative elasticities of substitution.

Elasticity of substitution, σ		*.50*	*.25*	*.15*
Supply constraints		*(same as base case)*	*(base case)*	*(with additional supply constraints)*
Aggregate consumption in 2010 ($ billion at 1975 price level)	With nuclear	3,566	3,496	3,245
	No nuclear	3,552	3,387	2,799
Reduction in consumption in 2010	($ billion)	14	109	446
	(%)	.4%	3.1%	13.7%
Public discount rate:	Present value of reduction in consumption, 1975–2050 ($ billion)			
5%		293	595	3,089
10%		36	77	456
Case identification (see tables below)		2,7	1,6	5,10

the total losses in aggregate consumption are $109 billion, or 3.1 percent of aggregate consumption in 2010. (See middle column of Table 2.) Adding over *all* years from 1975 through 2050, the present value of these losses would be $77 or $595 billions—depending upon whether public decisions are to be based on a 10 percent or a 5 percent discount rate. Note that the macroeconomic losses are low initially, and that they rise rapidly after the year 2000. This is why a 5 percent discount rate implies that the *present value* of these losses would be far higher than twice those associated with a 10 percent rate.

The base case implies that although a "no nuclear" policy would have negligible macroeconomic effects, there *would* be impacts throughout the energy sector. In the year 2010, for example, there would have to be sufficient price-induced conservation so that primary energy consumption would be 134.4 instead of 158.1 quads. (See Figure 3 and also Table 3.) Electricity generation would be affected by a still larger amount (5.2 instead of 8.1

Table 3. Energy demand comparison—four alternative cases.

Nuclear energy policy		*With nuclear*		*No nuclear*	
Case identification number		1 (base)	2	6	7
Elasticity of substitution, σ		.25	.50	.25	.50
	1975	71.0	71.0	71.0	71.0
Total primary	2000	126.2	107.4	118.1	104.4
energy consumption	2010	158.1	125.6	134.4	109.3
(quads)	2020	192.3	150.9	165.4	118.5
	1975	2.0	2.0	2.0	2.0
Electricity	2000	5.6	4.3	4.3	4.1
generation	2010	8.1	5.8	5.2	4.5
(trillion kwh)	2020	9.9	7.6	6.9	4.8

trillion kwh in 2010). In general, energy costs do not exceed 10 percent of gross output. (See the macro variable EC in Table 4.)

The preceding results have been based upon $\sigma = .25$ as the best available estimate of the elasticity of substitution between primary energy and other economic inputs. With a higher elasticity of substitution, the macroeconomic impact of a "no nuclear" policy would be lower than indicated by the base case. If, for example, it is assumed that $\sigma = .50$, the macroeconomic impact is barley discernible until 2020—amounting to $71 billion in that year. (See bottom rows, Table 4, cases 2, 7.) In this case, energy demands grow more slowly (Table 3), oil and gas resources would become exhausted less rapidly, and there would be more time for a transition to future high-cost alternative energy sources. Price-induced conservation would solve our energy problems for the next fifty years—*if* it turns out that $\sigma = .50$ or higher.

7. IMPLICATIONS OF A LOW ELASTICITY OF SUBSTITUTION

If the elasticity of substitution is sufficiently low, there could be major economic impacts from tight energy supplies. To illustrate this possibility, we have computed one set of cases in which the elasticity parameter is set at .15 (instead of .25). In addition, a series of reductions (or cost increases) have been made in the supplies of non-nuclear energy. The following *combination* of circumstances is improbable, but seems far more likely than some of the low-probability, high-consequence disasters that have been discussed in connection with nuclear energy:

(a) the upper bound upon coal consumption is reduced to 25 quads[15]

Table 4. Macroeconomic results—ten alternative cases.

Nuclear energy policy	*Year*	*Macro variable*	*Base* (σ=.25, δ=10%)	σ=.5	δ=8%	δ=12%	σ=.15, *etc.*
			(unit: $\$10^{12}$/year—at 1975 price level)				
			Case 1	Case 2	Case 3	Case 4	Case 5
With nuclear	2000	C	2.763	2.795	2.820	2.709	2.757
		I	.497	.520	.581	.436	.430
		EC	.254	.195	.269	.240	.273
		Y	3.514	3.510	3.670	3.385	3.459
	2010	C	3.496	3.566	3.591	3.411	3.245
		I	.574	.603	.677	.497	.491
		EC	.376	.290	.405	.352	.483
		Y	4.446	4.459	4.673	4.260	44.220
	2020	C	4.244	4.375	4.382	4.121	3.869
		I	.659	.681	.776	.575	.558
		EC	.517	.346	.550	.491	.723
		Y	5.420	5.402	5.708	5.187	5.150
			Case 6	Case 7	Case 8	Case 9	Case 10
No nuclear	2000	C	2.757	2.790	2.811	2.705	2.625
		I	.465	.515	.545	.407	.231
		EC	.250	.195	.265	.237	.296
		Y	3.473	3.500	3.621	3.350	3.152
	2010	C	3.387	3.552	3.477	3.305	2.799
		I	.553	.583	.647	.484	.351
		EC	.395	.262	.424	.372	.629
		Y	4.335	4.397	4.548	4.161	3.780
	2020	C	4.109	4.304	4.236	3.995	3.218
		I	.641	.657	.758	.554	.396
		EC	.562	.336	.599	.527	.896
		Y	5.311	5.297	5.592	5.077	4.509
Consumption loss	2000		.006	.005	.009	.004	.132
	2010		.109	.014	.114	.106	.446
	2020		.135	.071	.146	.126	.651

Note: Due to rounding, totals need not add.

in the year 2000 (instead of 40 quads) and to an asymptote of 40 (instead of 50 quads);

(b) no shale oil resources are developed because of air and water quality limits;

(c) the introduction date for large-scale direct solar electricity is delayed from the year 2000 to 2020; and

(d) AES costs are equivalent to oil and gas at \$8 instead of \$5 per million BTU.[16]

Even under these unfavorable circumstances, aggregate economic activity would be only moderately affected *if* nuclear energy is available. In the year 2000, for example, there would be virtually no difference in total consumption.

With a "no nuclear" policy, however, the economic losses would be enormous. They would amount to \$132 billion in the year 2000, and would grow to \$446 billion in 2010. (See rightmost column in Table 2.) This would mean a 13.7 percent reduction of consumption in the latter year. Energy costs would then amount to \$629 billion or 16.6 percent of gross output. (See Table 4.)

8. DISCOUNT RATES AND UTILITY FUNCTIONS

On philosophical grounds, there is no easy way to determine a numerical value for the key savings parameter, the utility discount rate δ. Ramsey, Harrod and others have argued that this rate *ought* to be zero.[17] With a zero value for δ, however, it is known that there would be convergence difficulties and "horizon effects" with a logarithmic utility function.[18] Rather than employ a less intuitive form of such a function,[19] it appeared more satisfactory to adopt a positive social discount rate for this initial set of experiments with ETA-MACRO.

With a logarithmic utility function, the Ramsey rule (i.e., the first-order condition for an optimal growth trajectory) takes on the following special form:

optimal growth rate of consumption	=	marginal physical productivity of capital	−	utility discount rate

$$\frac{\Delta C}{C} = \frac{\partial Y}{\partial K} - \delta. \qquad (9)$$

From the production function (2), it was estimated that the gross return on capital was 13.3% per year in 1970. Thus, if δ is set at a much lower value than 10%, equation (9) implies that it would be optimal to depress consumption immediately in order to permit a rapid buildup of investment and capital stocks. This would gradually reduce the marginal productivity of physical capital, and eventually there would be a transition to a trajectory in which it is again optimal for consumption to grow at the same rate as the labor force (measured in efficiency units).

Equation (9) provides considerable insight into the numerical experiments with alternative discount rates. As δ is reduced from 12 percent to 10 percent to 8 percent per year, note that there is a sharp increase in the 1985 investment rate. (Table 5.)[20] Thus, the lower the value of δ, the more serious would become the "crowding-out" problem in the capital markets, and the more serious might become the inflationary consequences. In addition to the results shown here, we have also experimented with a 5 percent utility discount rate. This leads to a rather implausible investment stepup for 1985, and it does not appear worthwhile to report the numerical results in detail.

Table 5. Implications of alternative discount rates.

Utility discount rate, δ *Case number*		*8%* *3*	*10%* *1* (*base case*)	*12%* *4*
Macro variable	Year	(unit: $\$10^{12}$/year—at 1975 price level)		
	1975	.200	.200	.200
I = investment	1985	.373	.319	.280
	2000	.581	.497	.436
	2020	.776	.659	.575
	1975	1.300	1.300	1.300
	1980	1.438	1.475	1.504
C = consumption	1985	1.760	1.754	1.742
	2000	2.820	2.763	2.709
	2020	4.382	4.244	4.121
			(unit: %)	
	1975	13%	13%	13%
$\frac{I}{I+C}$ = investment—	1985	17	15	14
GNP ratio	2000	17	15	14
	2020	15	13	12
			(unit: %/year)	
$\frac{\Delta C}{C}$ = annual growth	1975–85	3.1	3.0	3.0
rate of	1985–2000	3.2	3.1	3.0
consumption	2000–2020	2.2	2.2	2.1

Whatever may be one's philosophical views about utility functions, discount rates, etc., it does *not* appear that they have a major impact upon the calculation of the economic losses associated with a "no nuclear" energy policy. From Table 4, recall that these losses range from $106 to $114 billion in the year 2010—as δ varies from 8 percent to 12 percent per year. Given the other uncertainties involved in long-term energy projections, this difference appears negligible.

These discount rate experiments suggest that many energy policy results may turn out to be invariant to the rather arbitrary choice of the macro-economic growth criterion. All that seems needed is sufficient discounting—or sufficient curvature of the utility function—so that aggregate investment is not stepped up too rapidly over the short run. For U.S. energy policy, the crucial econometric parameters are the elasticity of substitution between energy and other inputs—together with the rate of growth of the labor force and of its productivity at constant energy prices. These are the parameters that appear to dominate the long-term growth picture.

* * *

Appendix A and excerpts from Appendix C are reproduced on the following pages. A complete set of Appendices B and C is available upon request to the author.

APPENDIX A. STRUCTURE OF THE ETA SUBMODEL

1. User-specified information

To embed ETA within ETA-MACRO, the user inserts energy sector information into the coding format of Appendix B. For each run, six pages must first be filled out:

page B-1: discount rate (δ), elasticity of substitution (σ), electricity value share (β), capital value share (α), and 1970 "reference price" of crude oil (p);
page B-2: indices of GNP—at constant energy prices of 1970
page B-3: upper bounds (if any) upon total coal consumption
page B-4: cumulative quantity versus selling price of petroleum and natural gas resources
page B-5: cumulative quantity versus selling price of coal resources
page B-6: cumulative quantity versus selling price of uranium resources

In addition, there is a "technology list" containing one page for each supply or conservation activity. From Table 1, recall that there are sixteen such activities in this version of ETA. Each page must contain the following information on the input and output *coefficients* per unit of activity j:

A(ELEC): annual electrical output
A(NELE): annual nonelectrical output (liquids and gases)
A(PETG): annual input of petroleum or natural gas
A(COAL): annual input of coal
A(NATU): annual input of natural uranium
A(PLUT): annual input (or output) of plutonium
A(SWUS): annual input of separative work units for uranium enrichment
cur: annual equivalent of current account costs—fuel, operating and possibly capital charges
cap: initial capital costs *not* included above
B(NATU): initial inventory of natural uranium
B(SWUS): initial inventory of separative work units for uranium enrichment.
B(ELIN): total investment costs for electricity generation, transmission and distribution
B(NEIN): total investment costs for nonelectric energy supply or conservation activities

The user is given the option of defining which of these coefficients remain constants over time, and which ones are time-dependent. An activity may be readily included in some cases and excluded from others within a single run. For each individual time period, the level of the jth activity may be fixed in advance, or it may be subject to either an upper or lower bound.

2. Algebraic formulation of constraints

The classes of activity columns and constraint rows are listed in Tables 1 and 2, respectively. The following subscripts and notation are employed:

i:	input or output row identification
j:	activity identification (supply or conservation technologies)
J:	number of technologies for which capacity equations must be written explicitly
JMAX:	total number of technologies
r:	fuel resources
l:	cost level for fuel resources
LMAX:	total number of cost levels for all fuel resources
t:	time period

The constraints are formulated as follows. Parameters are barred above, and unknowns are unbarred:

CP(j)(t)—capacities

$$PC(j)(t) = PC(j)(t-5) + 5[DP(j)(t) - DP(j)(t-30)]$$

(all t; all j, except for PETG and where "no" is indicated on technology list)

N.B. By definition, all activity levels are zero for $t \neq 0, 5, .. 75$.

Table 1. Activity column summary—ETA submodel.

		Number of columns
DM (i) (t)	demand for item i = ELEC, NELE respectively*	16·2
PC (j) (t)	production of energy at predetermined capacity factor, technology j, period t	16 JMAX
DP (j) (t)	capacity increments, annual rates, technology j, period t	16 J
Q (r) (*l*) (t)	quantity of fuel resource r, cost level *l* required in time period t	16 LMAX
CSPLUT (t)	cumulative stockpile of plutonium available at beginning of time period t	16
ENCOST (t)	energy sector costs, period t*	16
	Total number of columns	16 (J + JMAX + LMAX + 4)

*Here we employ an eight-character notation for all unknowns. The original macro variables are translated as follows:

E(t) = DMELEC(t)
N(t) = DMNELE(t)
EC(t) = ENCOST(t)

Table 2. Constraint row summary—ETA submodel

		Number of rows
CP(j)(t)	capacity available	16 J
RQ(i)(t)	demand rows (i = ELEC, NELE)	16·2
RQ(r)(t)	requirement rows (r = PETG, COAL, NATU, SWUS)	16·4
AV(r)(*l*)	availability of depletable natural resources (r = PETG, COAL, NATU) at cost level *l*,	LMAX
SM(r)(t)	cumulative fuel resource requirements (r = PETG, COAL, NATU, PLUT)	16·4
IV(i)(t)	annual investment rates	15·2
CL(i)(t)	coal constraints	16·2 + 9
COSTEN(t)	energy costs	16
	Total number of rows*	247 + 16 J + LMAX

*This total includes 94 "non-binding" rows employed for various report-writing steps, but excludes upper and lower bounds on individual unknowns.

For new technologies such as the fast breeder reactor (FBRX) or coal-based synthetic fuels (SYNF), the user will typically fix the activity level PC(j)(t) at zero for time periods 0, 5, ... T-5, and will specify upper bounds $\overline{\mathrm{PC(j)(t)}}$ for periods T, T + 5, ... 75. Occasionally, it turns out that these upper bounds are *not* binding constraints during period T or T + 5, but that they are thereafter. To prevent too rapid a rate of introduction during the periods following T, we therefore impose the following additional upper bounds:

$$5\,\mathrm{DP(j)(t)} \le \overline{\mathrm{PC(j)(t)}} - \overline{\mathrm{PC(j)(t-5)}} \qquad (\mathrm{t} = \mathrm{T}+5, \mathrm{T}+10, \ldots \mathrm{T}+25).$$

CP(PETG)(t) – petroleum production decline curve

$$.2\,\mathrm{PC(PETG)(t)} = \sum_{\tau=0}^{5} \mathrm{DP(PETG)(t}-\tau) + \sum_{\tau=10}^{35} (.93)^{\tau-5}\,\mathrm{DP(PETG)(t}-\tau) \qquad (\text{all t})$$

RQ(i)(t) – demand rows (i = ELEC, NELE; all t)

$$\sum_{\mathrm{j}} \overline{\mathrm{A(ELEC)(j)}}\,\mathrm{PC(j)(t)} \ge \mathrm{DM(ELEC)(t)}$$

$$\sum_{\mathrm{j}} \overline{\mathrm{A(NELE)(j)}}\,\mathrm{PC(j)(t)} \ge 100\,\mathrm{DM(NELE)(t)}$$

N.B. To avoid scaling difficulties in the nonlinear programming routine the activities DMELEC(t) are expressed in trillion KWH, and DMNELE(t) in *hundreds* of quads per year—hence the factor of 100 that is associated with the latter variable.

RQ(r)(t) – natural resource requirements

$$\sum_{l=1}^{9} \mathrm{Q(r)}(l)\mathrm{(t)} \geq \sum_{j} \overline{\mathrm{A(r)(j)}}\,\mathrm{PC(j)(t)} + \sum_{j} \overline{\mathrm{B(r)(j)}}[\mathrm{DP(j)(t+5)} - \mathrm{DP(j)(t-30)}]$$

(all t; r = PETG, COAL, NATU)

RQ(SWUS)(t) – enrichment requirements

$$\sum_{j} \overline{\mathrm{A(SWUS)(j)}}\,\mathrm{PC(j)(t)} + \sum_{j} \overline{\mathrm{B(SWUS)(j)}}[\mathrm{DP(j)(t+5)} - \mathrm{DP(j)(t-30)}]$$

(all t; non-binding)

AV(r)(l)—availability of natural resources

$$\sum_{t=0}^{75} 5(\mathrm{Q(r)}(l)\mathrm{(t)} \leq \overline{\mathrm{CA(r)}(l)} - \overline{\mathrm{CA(r)}(l} - 1)$$

(r = PETG, COAL, NATU; $l = 1 \ldots 9$; by definition, $\overline{\mathrm{CA(r)(0)}} \equiv 0$.)

SM(r)(t) – cumulative resource requirements

$$\sum_{l=1}^{9} \sum_{\tau=0}^{t} 5\mathrm{Q(r)}(l)(\tau) \qquad \text{(non-binding)}$$

(all t; r = PETG, COAL, NATU)

SM(PLUT)(t) – cumulative plutonium stockpile

$$\begin{aligned} &.20\,\mathrm{CSPLUT(t)} - .19\,\mathrm{CSPLUT(t-5)} = \\ &-\sum_{j} \overline{\mathrm{APLUT(j)}}\,\mathrm{PC(j)(t-5)} \\ &-\sum_{j} \overline{\mathrm{BPLUT(j)}}[\mathrm{DP(j)(t)} - \mathrm{DP(j)(t-30)}]. \end{aligned}$$

(all t)

IV(i)(t) – annual investment rates

$$\sum_{j} \overline{\mathrm{B(i)(j)}}\,\mathrm{DP(j)(t)}$$

(i = ELIN, NEIN; t ≥ 5; non-binding)

CL(i)(t) – coal limits

i = TOTL (approximates the effect of limits upon the total use of coal because of health, environmental and global climatological consequences)

$\sum_j \overline{A(COAL(j)} PC(j)(t) \leq$ RHS constants (as listed on input page B-3)

(all t)

---

i = DLOC (applies if and only if COLL appears on technology list; approximates the effect of regional differences in coal costs; restricts low-cost regions to no more than 50 percent of total new generating capacity)

$$\overline{A(ELEC)(COLL)}\, DP(COLL)(t) \leq \sum_{j \dagger COLL} \max [0, \overline{AELEC(j)}]\, DP(j)(t). \quad \text{(all t)}$$

---

i = DMIN (approximates the economic advantages of coal for intermediate and peaking duty; restricts coal to be at least 40 percent of total new generating capacity until the time period when advanced storage technologies become available)

$$THETA(t) \Sigma \max [0, \overline{AELEC(j)}]\, DP(j)(t)$$
$$\leq [1 - THETA(t)] \sum_{j \varepsilon SET} A(ELEC)(j)\, DP(j)(t)$$

where jεSET if A (ALEC(j) > 0 and A(COAL)(j) > 0.

t =	THETA(t) =
5, . . 25	.40
30	.30
35	.20
40	.10

COSTEN(t) – energy costs

$$1000\, ENCOST(t) = \sum_j \overline{cur(j)}\, PC(j)(t) + \sum_{r,l} PR(r)(l)Q(r)(l)(t) \quad \text{(all t)}$$

Notes:

(1) Activities ENCOST(t) are measured in $\$10^{12}$ to avoid numerical scaling difficulties. The right-hand terms are measured in $\$10^9$—hence the factor of 1000 on the left hand.

(2) To embed ETA within ETA-MACRO, all capital costs within the energy sector must be levelized at the fixed charge rate prevailing within that industry, and then included in the cost coefficient cur(j). We do *not* employ the coefficients cap(j) here.

ETA-MACRO---CASE: 1(ETA-MACRO BASE CASE

	1985	1990	2000	2010	2020	2030
1. DOMESTIC ENERGY PRODUCTION, BY SOURCE, NET OF EXPORTS (QUADS)	66.6	78.1	105.1	142.3	185.6	229.5
1.1 COAL, TOTAL:	17.9	21.2	28.7	45.0	47.7	49.0
A. FOR ELECTRICITY GENERATION	12.4	15.1	21.3	24.5	18.5	9.4
B. FOR SYNTHETIC FUELS	----	0.0	0.0	11.5	18.2	26.2
C. DIRECT USE AND OTHER	5.5	6.1	7.4	9.0	11.0	13.4
1.2 NATURAL GAS, OIL AND NGL	36.5	39.8	42.1	31.6	13.4	5.0
1.3 NUCLEAR, TOTAL:	8.3	12.7	27.1	48.1	70.1	95.3
LWR	8.3	12.7	27.1	46.5	47.7	33.3
FBR	----	----	0.0	1.6	22.3	62.0
1.4 SOLAR ELECTRIC	----	----	0.0	0.0	0.0	0.0
1.5 SHALE OIL	----	----	1.5	5.6	9.6	11.5
1.6 HYDRO, GEOTHERMAL, ETC.	3.9	4.4	5.7	7.2	9.3	11.9
1.7 NON-ELECTRIC AES	0.0	0.0	0.0	4.9	35.6	56.8
2. OIL AND GAS IMPORTS (QUADS)	18.2	19.9	21.1	15.8	6.7	2.5
3. TOTAL ENERGY CONSUMPTION (QUADS)	84.9	98.0	126.2	158.1	192.3	232.0
4. ELECTRICITY GENERATION (TRILLION KWH)	2.9	3.6	5.6	9.1	9.9	11.7
4.1 % OIL AND GAS	14.7	9.0	1.9	0.0	0.0	0.0
4.2 % COAL	43.6	43.5	39.7	31.8	19.7	9.4
4.3 % NUCLEAR	28.4	35.3	48.3	59.3	70.9	81.4
4.4 % OTHER	13.3	12.2	10.1	8.9	9.4	10.2
5. URANIUM CONSUMPTION (MILLION TONS OF U308, CUMULATIVE FROM 1975)	0.2	0.4	1.0	2.1	3.5	3.7
6. DOMESTIC PRICES:						
6.1 COAL ($/MM BTU)	0.7	0.7	0.7	1.5	1.5	1.5
6.2 OIL AND GAS ($/MM BTU)	2.2	2.4	3.3	5.0	5.0	5.0
6.3 ELECTRICITY (MILLS/KWH)	21.7	22.0	22.1	24.2	24.6	25.8
6.4 URANIUM ($/LB. U308)	30.7	31.3	34.4	44.5	75.7	141.2

ETA-MACRO---CASE: 6(CASE 1 BUT NO NUCLEAR

	1985	1990	2000	2010	2020	2030
1. DOMESTIC ENERGY PRODUCTION, BY SOURCE, NET OF EXPORTS (QUADS)	65.7	76.8	95.0	119.4	159.1	196.2
1.1 COAL, TOTAL:	23.6	30.9	40.0	45.0	47.7	49.0
A. FOR ELECTRICITY GENERATION	18.1	24.8	32.6	36.0	36.7	35.6
B. FOR SYNTHETIC FUELS	----	0.0	0.0	0.0	0.0	0.0
C. DIRECT USE AND OTHER	5.5	6.1	7.4	9.0	11.0	13.4
1.2 NATURAL GAS, OIL AND NGL	36.7	39.9	46.2	30.0	12.6	4.6
1.3 NUCLEAR, TOTAL:	1.5	1.5	1.5	0.0	0.0	0.0
LWR	1.5	1.5	1.5	0.0	0.0	0.0
FBR	----	----	0.0	0.0	0.0	0.0
1.4 SOLAR ELECTRIC	----	----	0.0	6.5	21.6	31.9
1.5 SHALE OIL	----	----	1.5	5.6	9.6	11.5
1.6 HYDRO, GEOTHERMAL, ETC.	3.9	4.4	5.7	7.2	9.3	11.9
1.7 NON-ELECTRIC AES	0.0	0.0	0.0	25.1	58.4	87.3
2. OIL AND GAS IMPORTS (QUADS)	18.3	20.0	23.1	15.0	6.3	2.3
3. TOTAL ENERGY CONSUMPTION (QUADS)	84.1	96.7	118.1	134.4	165.4	198.5
4. ELECTRICITY GENERATION (TRILLION KWH)	2.8	3.5	4.3	5.2	6.9	8.1
4.1 % OIL AND GAS	15.1	9.2	2.5	0.0	0.0	0.0
4.2 % COAL	65.8	73.8	80.4	73.4	55.6	46.1
4.3 % NUCLEAR	5.4	4.4	3.6	0.0	0.0	0.0
4.4 % OTHER	13.7	12.6	13.4	26.6	44.4	53.9
5. URANIUM CONSUMPTION (MILLION TONS OF U308, CUMULATIVE FROM 1975)	0.1	0.1	0.1	0.1	0.1	0.1
6. DOMESTIC PRICES:						
6.1 COAL ($/MM BTU)	0.8	0.8	1.9	3.2	2.4	2.9
6.2 OIL AND GAS ($/MM BTU)	2.2	2.4	3.3	5.0	5.0	5.0
6.3 ELECTRICITY (MILLS/KWH)	23.1	23.2	33.7	45.9	40.1	42.5
6.4 URANIUM ($/LB. U308)	30.0	30.0	30.0	30.0	30.0	30.0

FOOTNOTES

Presented at ORSA/TIMS Meeting, San Francisco, May 9–11, 1977. This material has been reproduced from a working paper volume with permission from Resources for the Future. The author is solely responsible for the views expressed here. This research was supported by a grant from the Electric Power Research Institute to Stanford University. Thanks go to Hung-po Chao, Robert Condap, Anthony De Andrade, Ray Squitieri, and Michael Saunders for their help in constructing ETA-MACRO. Helpful comments have been received from Kenneth Arrow, William Hogan, Lester Lave, and Stephen Peck.

1. For these long-run demand projections, we have supposed that conservation will be induced by permitting energy prices to rise toward their replacement costs (long-run marginal cost of production, including capital charges). An alternative—and perhaps more realistic interpretation—would be that consumer demands will continue to be stimulated by price controls that maintain artificially low-energy prices, but that these effects will just be offset by nonprice conservation measures such as auto efficiency standards and tax rebates for improved insulation.

2. See (12). Unlike the energy sector models utilized by the MRG, ETA-MACRO is specifically designed to analyze interdependence between the energy sector and the economy as a whole. In the MRG's single-sector models, the GNP growth rate is an exogenous datum, but in ETA-MACRO this rate is endogenous.

3. GNP percentage comparisons can lead to counterintuitive and possibly misleading estimates of the importance of new technologies. The history of the railroads affords an instructive parallel to today's debates over nuclear power. Most historians have believed that the railroads were a major factor in the nineteenth century development of the U.S. economy. Fogel (4), however, has calculated that if the entire railroad industry had *not* been developed, the loss would have been only 5 percent of the U.S. GNP in 1890. Also see the critical review of this estimate in P. A. David (3, ch. 6).

4. For the concepts and terminology of macroeconomic production functions, neoclassical growth theory, etc., see, e.g., Allen (1).

5. Let p denote the "reference" price of nonelectric energy in the equilibrium year 1970. Then a first-order optimality condition implies that:

$$\frac{\partial Y}{\partial N} = Y^{1-\rho} E^{\rho\beta} N^{\rho(1-\beta)-1}(1-\beta)b = p = \$.80/10^6\,\text{BTU}.$$

Except for b, each element in the preceding equation is known. After solving for b, we employ the production function (2), defining the labor force index L = 1.0 in 1970. This permits us to solve as follows for a:

$$a\,K^{\alpha\rho} + b\,E^{\rho\beta} N^{\rho(1-\beta)} = Y^{\rho}.$$

6. For sensitivity tests, we shall also provide comparisons based upon $\sigma = .50$ and $\sigma = .15$.

7. Further details on these computations are available in a memorandum available upon request from the author.

8. These problems were solved by the reduced gradient algorithm described in Murtagh and Saunders (13). Among the special features of this program are:

(i) in-core implementation of the simplex method, using a sparse LU factorization of the basis; (ii) a numerically reliable implementation of a quasi-Newton method for optimization of the objective function within a given subspace; and (iii) an efficient line search routine.

Our specific problem contained 80 variables entering nonlinearly into the objective function. Altogether, there were 350 rows, 600 columns, and 2,000 nonzero matrix elements. On the Stanford Linear Accelerator's IBM 370/168, it required 90 seconds to solve one case from a cold start, and then approximately 30 seconds for each subsequent case.

9. From Figure 1, recall that the unknowns E(t), N(t) and EC(t) enter not only into equation (7), but also into the constraints of the energy sector sub-model ETA. For details, see Appendix A.

10. Caveat: The utility discount rate is not to be confused with the cost of capital or its marginal productivity within the private sector.

11. The logarithm implies that marginal utility is always positive, but is a diminishing function of the aggregate level of consumption. Daniel Bernouilli used a logarithmic form for the utility analysis of risk aversion during the eighteenth century.

12. To allow for the health, environmental and global climatological consequences of extensive use of coal, the MRG defined an upper bound upon the annual rate of coal consumption for all purposes taken together—coal-fired electricity, direct uses and synthetic fuels. This upper bound takes the form of a logistic curve defined by three points: the actual 1970 value of 13 quads, a projected value of 40 quads for the year 2000, and a long-term asymptote of 50 quads. This overall constraint is defined in Appendix A through the rows CLTOTL(t).

For ETA, we have included additional constraints upon the use of coal for electricity generation. This is done to approximate the effects of regional differences in the cost of coal delivered to electric utilities—and also to allow for the advantages of coal in intermediate and peaking duty until the time period when advanced storage technologies become available. ETA therefore includes upper bounds on the use of coal in low-cost-regions, and it also includes lower bounds to allow for the intermediate and peaking requirements. These upper and lower bounds are denoted respectively, CLDLOC(t) and CLDMIN(t) in Appendix A.

13. One other parameter change has led to an increase in electricity demands. In the MRG runs of ETA, the electricity value share parameter was set at .333. Here the value share $\beta = .400$. This is a rough adjustment to allow for Sergio Granville's recent econometric work suggesting that the elasticity of substitution between electric and nonelectric energy may be as high as 1.25 or 1.50.

14. For these purposes, we interpret a "no nuclear" policy to mean that no net new nuclear capacity will be installed—over and above the amounts operating in 1975. Moreover, these plants will be retired at the end of their thirty-year service life. In the technology list (Table 1 above), this option is described as LWRB.

15. Recall that the 1970 level was 13 quads, and that coal consumption had not increased during the period since World War II.

16. The AES (alternative energy system) was postulated by the CONAES Modeling Resource Group as a backstop technology for producing clean nonelectric energy. Solar heating and cooling would be one example of such a backstop. Other possibilities would include *in situ* shale oil retorting, biomass conversion, solar-generated hydrogen, and unconventional petroleum resources.

Under base case assumptions, the AES does not become a significant energy source until 2010 or thereabouts. (See Appendix C, line 1.7.) After 2020, however, the AES becomes the dominant source of new energy supplies. These AES introduction rates appear to be on the high side, and are probably more optimistic than those adopted here for better understood technologies such as synfuels and the breeder.

17. For a brief introduction to this controversy, see Koopmans (9).

18. See the "cake-eating" example in Gale (5).

19. The logarithmic function implies that 1.0 is the absolute value of the elasticity of *marginal* utility with respect to consumption. Had it been assumed that this elasticity

exceeds 1.0, we could have avoided the need for discounting, setting $\delta = 0$. For a readable exposition of these issues—including the problems of convergence with an infinite planning horizon—see Chakravarty (2, esp. chs. 1–4.).

20. One additional feature should be noted in Table 5. The 1975–1985 rate of GNP growth (3.3 percent for the base case) lies well below the *potential* rate of 4.0 percent specified for the labor force index. A portion of this gap may be connected with aberrations in the initial conditions for 1975, but it also seems likely that a significant portion is attributable to the economic drag exerted by rising energy prices.

REFERENCES

1. Allen, R. G. D., *Macro-Economic Theory*. Macmillan, New York, 1968.
2. Chakravarty, S., *Capital and Development Planning*. MIT Press, Cambridge, Mass., 1969.
3. David, P. A., *Technical Choice, Innovation and Economic Growth*. Cambridge University Press, 1965, ch. 6.
4. Fogel, R. W., *Railroads and American Economic Development*. Johns Hopkins University Press, Baltimore, 1964.
5. Gale, D., "On Optimal Development in a Multi-Sector Economy," *Review of Economic Studies*, vol. 34 (1967) 1.
6. Hogan, W., and A. Manne, "Energy-Economy Interactions: the Fable of the Elephant and the Rabbit?" Energy Modeling Forum, Working Paper, EMF 1.3, Stanford University, June 1977.
7. Hudson, E., and D. Jorgenson, "U.S. Energy Policy and Economic Growth, 1975–2000," *Bell Journal of Economics and Management Science*, vol. 5 (1974) 2.
8. Keeny, S. M., Jr., *et al.*, *Nuclear Power Issues and Choices*. Ballinger Press, Cambridge, Mass., 1977.
9. Koopmans, T. C., "Ways of Looking at Future Economic Growth, Resource and Energy Use," in M. Macrakis (ed.), *Energy: Demand, Conservation and Institutional Problems*. MIT Press, Cambridge, Mass., 1974, ch. 1.
10. Manne, A. S., "ETA: A Model for Energy Technology Assessment," *Bell Journal of Economics*, vol. 7 (1976) 2.
11. Mitchell, E. J., *U.S. Energy Policy: A Primer*. American Enterprise Institute for Public Policy Research, Washington, D.C., 1974, p. 79.
12. Modeling Resource Group, Committee on Nuclear and Alternative Energy Systems (CONAES), National Research Council, Washington, D. C., 1977.
13. Murtagh, B. A., and M. A. Saunders, "MINOS: A Large-Scale Nonlinear Programming System," Technical Report SOL 77–9, Department of Operations Research, Stanford University, February 1977.
14. Starr, C., and E. Zebroski, *Nuclear Power and Weapons Proliferation*. Electric Power Research Institute, Palo Alto, Calif., 1977.

THE BAUXITE CARTEL IN THE NEW INTERNATIONAL ECONOMIC ORDER

Harold J. Barnett, WASHINGTON UNIVERSITY, ST. LOUIS

NEW ECONOMIC ORDER

A new international order has been emerging since World War II, especially in more recent years. In the less-developed countries (LDCs) the world over, by coups and revolutions, and less frequently by democratic voting processes, new governments, usually with right or radical left (mostly left) views, have taken over—in Africa, Asia, and Latin America. Resident colonial powers have voluntarily departed or have been evicted. The super powers have usually not interfered greatly with this process, primarily because of stalemate and detente.

New nations joined the UN, and began to exercise their UN powers of

Advances in the Economics of Energy and Resources, Vol. 2, pp. 235–250.
Copyright © 1979 by JAI Press, Inc.
All rights of reproduction in any form reserved.
ISBN 0-89232-079-6

expression and to vote. They formed coalitions under such names as Third World and Non-aligned Nations. They began to express themselves in concert, in the UN Assembly or at other conferences. They represent a vast majority of the UN General Assembly. Their basic thrust is that they want an end to the present international order of unequal political and economic power among nations. They want a redistribution of power and wealth. Their justification is suggested by the terms they use, such as fairness, justice, equity and morality. The existing evil is suggested by such terms as power, imperialism, colonialism, exploitation, and unequal exchange.

During 1973–1974, the New Economic Order (NEO), the major element of the New International Order, received great impetus. It was given specificity, form, and force in the exercise of strong cartel power by OPEC, the cartel Organization of the Petroleum Exporting Countries. Here, starkly visible to the Less Developed Countries, was evidence that they were not engaged merely in wishful thinking. They could demand and, by political and economic action with respect to their natural resources, force the redistribution of economic power and wealth to which they aspired.

DECLARATIONS

In political concert, following the demonstration of OPEC power, the LDCs and the DCs who followed them gleefully or unhappily acted in the General Assembly. They adopted the Charter of Economic Rights and Duties of States Resolution 3281, 29th UN General Assembly, December 12, 1974 by a vote of 120 in favor and six against, with ten abstentions. I quote, emphasizing resources provisions:

> ... it is a fundamental purpose of this Charter to promote the establishment of the new international economic order .... Every State has and shall freely exercise full permanent sovereignty, including possession, use and disposal, over all its wealth, natural resources and economic activities. Each State has the right: (a) To regulate and exercise authority over foreign investment.... (b) To regulate and supervise the activities of transnational corporations.... (c) To nationalize, expropriate, or transfer ownership of foreign property, in which case appropriate compensation should be paid by the State adopting such measures, taking into account its relevant laws and regulations and all circumstances that the State considers pertinent. In any case where the question of compensation gives rise to a controversy, it shall be settled under the domestic laws of the nationalizing State and by its tribunals.... All States have the right to associate in organizations of primary commodity producers.... Correspondingly all States have the duty

> to respect that right.... States shall take measures aimed at securing additional benefits for the international trade of developing countries... a balance more favorable to developing countries... measures designed to attain stable, equitable, and remunerative prices for primary products.

The "Dakar Declaration" of the February 3–8, 1975, Conference of Developing Countries on Raw Materials, states (in UN–ECOSOC document E/AC. 62/6 of April 15, 1975):

> ... 1. The present structure of international trade, which had its origins in imperialist and colonialist exploitation, and which has continued in force up to the present day ... needs to be replaced by a new international economic order... A powerful weapon which the developing countries can use to change this state of affairs is to defend their natural resources... by combining their forces.... 22. The inequities and weaknesses of the present economic system are particularly glaring in the conduct of world trade in raw materials. Those who control the levers of the price mechanism have successfully denied to the producers of a number of raw materials their due profit from their labor and from their natural endowment, while they themselves have continued to make excessive profits by charging high prices for the finished products.

There have been further declarations, meetings, and institutions. At present there are four Conferences on International Economic Cooperation (CIEC) meeting in Paris and elsewhere, until the end of 1976, concerned with energy, raw materials, finance, and development.

The Nairobi Conference completed in May 1976 has had statements that natural resources are to be used to redistribute the world's wealth and income, promote justice and equity, bring fairness to the conduct of economic trade, etc. The U.S. Secretary of State has proposed an International Resources Bank to provide investment funds to the LDCs for their natural resources. Other measures are to provide for moratoria or cancellation of LDC debts, which have shot upward due to fivefold increases in oil prices and inflation; to finance buffer stocks, in order to improve markets for LDC exporters of raw materials; and to transfer technology to the LDCs.

Let us consider the simple magnitude of the OPEC achievement. In 1974, the OPEC LDCs produced and exported about 11 billion barrels of oil. At this output level and the prices and terms of 1972, the gross royalties, taxes, levies, and wage payments to the LDCs—i.e., the OPEC national gross benefits—totaled perhaps $11 billion. (All dollar figures in the paper are U.S.) In 1974, the actual figure exceeded $100 billion. The OPEC nations simply raised net selling prices of oil from about \$2 to about \$12 and retained the gain.

Moreover, the OPEC members have used small amounts of their revenue gains to buy out the oil companies at net book value or less, far, far below market value. At this point, the multinational companies (MNCs) produce and buy the oil at about \$12 a barrel, for a profit much below \$1 a barrel. The remainder of the sales revenues of more than \$11 a barrel flows to the oil exporting nations. A figure of less than \$1 a barrel covers the domestic cost of exploration, development, and production to the OPEC members on their nationalized properties.

The tenfold increase in oil revenues which flow to the respective governments could radically improve capital investments in infrastructure, producer durables, education, military strength, and other things which the LDCs desire.

It is this demonstration of "raw materials power" which is exciting and driving the governments of the materials-exporting LDCs. In a single year, following the 1973 and 1974 oil price rises, some of the oil producing LDC nations have jumped over most of the world to join the rich nations.

IBA FORMATION

In this context of OPEC example and New Economic Order thrust, commodity cartels have burst into being or been renewed: copper, coffee, tin, bauxite, and others. Among these is the International Bauxite Asociation (IBA), which is headquartered in Jamaica. We were concerned here with the impact on aluminum and bauxite markets of the IBA cartel and its members on the developed countries. The IBA was formed in 1974 by Jamaica, Surinam, Guinea, Guyana, Australia, Sierra Leone and Yugoslavia. Subsequently it has enrolled the Dominican Republic, Haiti, Ghana, and Indonesia. The IBA and it members have the objective of controlling raw material supplies and greatly increasing their national returns.

The cartel members produce 85 percent of the world's market supply of tropical bauxite, the ore from which aluminum is currently produced. A two-stage process (bauxite ore to alumina, alumina to aluminum) reduces 4 or 5 tons of ore to 1 ton of aluminum. Other somewhat less concentrated aluminous ores, not now in significant use, are alternatives to tropical bauxite. Among these are anorthosite, alunite, dawsonite, kaolin clay, and laterite bauxite.

BAUXITE MARKET

The economic magnitudes in the bauxite and aluminum industries are not large enough for similar enormous transfers of wealth from the DCs to the

IBA cartel members. The IBA members produce about 60 million tons of the bauxite a year. This is about 85 percent of the supply to all of the market economies of the world. Before cartel interventions, the value (f.o.b.) was about $0.6 billion. This compared with the value of oil exported by OPEC countries before their 1973 cartel intervention of more than $20 billion, more than thirty times as large. Measured in any meaningful economic way, the OPEC output and economic opportunity dwarf the IBA. A view that an IBA cartel might produce effects comparable in magnitude to OPEC's achievement is inconsistent with the relative values of petroleum and bauxite.

However, if we put aside such grandiose notions, we are left with quite important questions on recent and prospective effects of the bauxite cartel.

BAUXITE CARTEL

We immediately rule out questions of economic warfare in the form of total embargoes, which would create major short-term supply disruptions of more than momentary nature to the developed countries. Embargoes would be economically senseless, indeed counterproductive.

The sources of 1974 bauxite production are shown in the following table. Australia would not engage in such behavior; it goes beyond the bounds of cartel manipulation to raise prices and revenues, and results in zero revenues.

Bauxite Production, 1974
(In million metric tons)

Cartel members:	
Australia	20.1
Jamaica	15.3
Surinam	7.1
Guinea	6.6
Guyana	3.2
Yugoslavia	2.4
Dominican Republic	1.4
Indonesia	1.3
Sierra Leone	.7
Haiti	.7
Ghana	.4
Subtotal	59.2

Other market countries:	
U.S.A.	2.0
France	2.9
Greece	3.0
India	1.1
Brazil	.7
Malaysia	.9
Other	.5
Subtotal	11.1
Subtotal market countries	70.3
Non-market countries	
USSR	4.3
China	.6
Hungary	2.8
Rumania	.3
Subtotal	8.0
Total world	78.3

Few if any LDC countries could delude themselves that an embargo, during which revenues fell to zero, could be economically useful. (During the oil "embargo" no one shut down output. It was only the Arab countries which reduced output slightly or moderately, and then *for political purposes*.)

It would be economically suicidal for any LDCs that did embargo. The aluminum companies have large stocks and so do the U.S. government and perhaps other governments. As discussed below there are alternative sources of supplies. These include increased supplies from other countries not engaged in embargo; increased output from U.S. bauxite reserves; emergency use of U.S. laterite bauxite; accelerated scrap collection; temporary substitutions for aluminum in common end product such as building materials, containers, wrappings. I do not say that industry and government officials should not undertake economical preparatory measures which are appropriate for estimated short-term supply contingencies, however unlikely. However, this is not my interest here.

CARTEL MEMBER ACTIONS

Jamaica emerged as the leader. Immediately following cartel formation in early 1974, and after several months of fruitless negotiations with reluctant companies, Jamaica took unilateral "new economic order" actions. It nationalized all land and also other aluminium company properties which were not currently being utilized in bauxite or alumina production or transportation; nationalized 51 percent of the bauxite-alumina companies; paid for these with low-interest serial notes, which are to be paid from company revenues; took control of bauxite prices; and imposed bauxite tax levies indexed to aluminum ingot prices in the United States, retroactive to the beginning of 1974. The other LDCs quickly followed Jamaica's lead in imposing a levy and some of them on other measures as well. This all occurred without formal actions by or agreements within IBA.

It should be noted that Jamaica is by far the largest supplier of bauxite (in form of bauxite and processed into alumina) to the United States. United States dependence on Jamaica and on the Caribbean countries is very great.

One major effect of these actions was a very large gain in revenues by the LDC cartel members, from imposition of the levy. If previously the tax was \$1 to \$2½ a ton, it has now gone to \$6 to \$15 a ton. If the levy gain as actually enforced is an average of \$8 a ton, then the LDC bauxite producers are receiving more than \$300 million a year from users of bauxite in the DCs. Beyond this, the LDCs have cut themselves in on a profit stream, which may total another \$50 million or possibly more.

BAUXITE MARKETS

The classic cartel problem is to raise prices and then amicably to share the sales. The cartel, formally or informally, or by following a leader, creates a higher price level. At this higher price level, some of the members would like to sell more than they are selling. Also, the non-cartel fringe suppliers would like to sell more. They can do this only by "shading" the price or levy or other terms. Such price shading, however, cuts into the sales volume of the other cartel members. If the cartel leaders accept the reductions in their sales, they preserve the cartel. If they cut prices to try to maintain volume and others also cut, a price war may develop and the cartel fails.

What has happened to prices and sales of bauxite is not easy to determine. There is no significant spot market for bauxite. The real-life situation is complicated by long-term contracts with quite complicated price formulas, penalties, bonuses, offsets, tax levies and incentives, etc. The bauxite market is a long-term contract market, in part because the alumina plants are specialized to one or another of two main types of bauxite, and are further adapted to characteristics of the particular bauxite within each type. Further, the bauxite real-life situation has the bauxite buyer in the position of owner-investor in the bauxite mine. The buyer who has sunk investment in a bauxite mine feels less free to buy cheaper elsewhere, particularly if he is holding serial notes whose interest and liquidation payments depend on his company sales. These complications make the economic data and analytical task more difficult than if the market were spot and prices easily visible.

BAUXITE PRICES

The price evidence which I have been able to find is that price shading relative to Jamaica's lead is occurring in several forms. The tax levy which Guinea has imposed is $6.50 a ton, relative to $15 a ton in the Jamaica formula. Guinea's bauxite is higher grade than Caribbean, but more distant from the United States. Guinea's levy represents price shading relative to the situation before Jamaica made its leadership move.

Australia, the most distant supplier, but the world's largest of lesser-grade bauxites, is price-shading in the same sense, since it has no levy (although one of its bauxite states, Queensland, has a levy of $.50 to $1.50 a ton). Guyana has two government-owned mines (nationalized from Alcan in 1971 and Reynolds in 1974). These do not pay a levy and there is evidence that the delivered prices to U.S. ports are lower. A recent percentage levy agreement announced by Alcoa and Surinam is 6 percent of the U.S.-realized aluminum price, as compared with $7\frac{1}{2}$ percent to 8 percent in the Jamaica formula. Indonesia appears to have shaded price to Japan. Jamaica's previously announced levy was 8 percent for this year, but there have been reports that Jamaica has been shading its own figure. So much for my detective work on prices.

SALES

What has happened to sales, present and future? Large actual reduction of output in some cartel members because of increases in others is not likely in bauxite. As noted, the companies have sunk investments in the several countries, their alumina plants are adapted to feed stock, and the practice in the aluminum industry is long-term supply contracts. However, small declines are possible. And, more important, we should see the cartel and its leaders in difficulty if we see sales and *supply expansion* (that is incremental supply and construction of new capacity) occurring in patterns different from the prior sales and existing capacity. Such evidence, in fact, is apparent.

The evidence on supply changes is flimsy but suggestive. In 1975, Jamaica appears to have suffered minor output decline, despite government efforts to prevent it by specifying minimum output for tax purposes. This was partly related to the recession and decline in aluminum demand that year, but also partly to disaffection by several of the aluminum companies over what they considered arbitrary government behavior. In Jamaica, the Revere plant was closed for a lengthy period (and may still be). Alcoa was shut down for a while, only partly for the ostensible reason of worker disorder in the plant.

CAPACITIES

What about capacity expansion? If this is *not* in the pattern of existing capacities, then it tells us that some countries are gaining, other countries are not, and the former are relatively undercutting the latter in some way. The only major expansions programed and planned from 1973 to 1979, and further expansion contemplated for those countries beyond that date, are described below (in metric tons of bauxite per annum):

Guinea:	
1973	4,055,000
1976	9,155,000
1979	12,155,000
Contemplated further	+ 13,000,000
Australia:	
1973	17,750,000
1976	20,065,000*
1979	20,065,000*
Contemplated for 1980	+ 11,200,000
Contemplated, no date	+ 2,500,000
Brazil:	
1973	805,000
1976	1,126,000
1979	4,376,000
Contemplated further	+ 6,000,000

*Production in 1974

In Jamaica (and the remainder of the Caribbean), which took the lead in the cartel and the new economic order impositions on the multinational companies (MNCs), no expansion is programed, planned, or contemplated. The favored areas are two other cartel members, Guinea and Australia, and a nonmember, Brazil. Even if we add the LDC cases of substantial contemplated expansion where there are no programs and plans actually in place, we only have (metric tons of bauxite per annum):

Sierra Leone:	contemplated	+ 1,200
Indonesia:	"	+ 2,000
British Solomon Islands:	"	+ 1,200

COMPANY INVESTMENT CRITERIA

Things seem not too ideal for the oligopoly leaders. We are led to believe that the MNCs view capacity expansion in Jamaica and perhaps other Caribbean countries unfavorably; and that they view capacity expansion in Guinea, Australia, and Brazil favorably. Can we construct an explanation, additional to price shading, of why and what this means relative to the cartel and its leading nations? We try to do this by comparing conditions for expansion of capacity and supply from Jamaica, on the one hand, with conditions for Guinea, Australia, and Brazil on the other.

Apart from price, the companies' interests in investment can be summarized in the following two propositions:

(a) *Low risk.* They want assurance or security—i.e., low risk—with respect to future bauxite supply and their investments.
(b) *Low investment.* All industrial companies are experiencing stringency in capital funds. The companies want to minimize drain on their capital for these new investments.

RISK CONCERNING SUPPLY AND INVESTMENTS

For the major companies and their very large bauxite needs there is no practical spot market. The concern for assured long-term supply rests on the enormous "downstream" investments and economic activity in alumina, aluminum, fabricating, etc. All of these pivot on the availability of the bauxite input. The smooth working of several thousand dollars of downstream investment and more than a thousand dollars of downstream value added depend upon each ton of aluminum content in ore, which costs less than $50 to produce.

The concern for lowering risk on investment is obvious. The U.S. companies are now highly concentrated in dependence on Jamaica for supply. They already have some hundreds of millions of dollars invested in Jamaica bauxite facilities. Recently the companies have learned that their foreign long-term supply contracts and investments from LDCs have a much higher risk element than they had thought before. If all foreign countries were equal in risk, they would now want to diversify their supply capacities and investments. Relatively, their bauxite investments in Guinea and Brazil are far smaller than in Jamaica. If there is risk of breakage, not all the eggs should be in one basket.

Apart from diversification, Guinea, Brazil, and Australia might be viewed as less risky than Jamaica. Politically, Jamaica is a young democracy in political unrest, tending toward Latin American style. Companies are uncertain where the political pressures from new economic order excitement and other forces will push the leaders. Guinea is one-party Marxist and Brazil one-party, military conservative. Both may be more stable politically than Jamaica, and less forced by political pressures in bauxite decisions. Australia, of course, is neither LDC nor a new economic order free swinger. It has discarded the Socialist government which was in power when Australia joined the cartel. In general, the MNCs view it as much safer in their supply planning and investment, second only to the United States itself.

Economically, Jamaica has already shown a belief that her economic interest is to squeeze more revenue from bauxite by limiting supply. The other three countries have much larger reserves of bauxite, iron, and other resources. They are embarked on decades-long programs for expansion in these varied minerals and other natural resources, dependent in significant degree on a continuing inflow of secure foreign capital. They depend on continuing MNC belief that the investment climate is favorable.

INVESTMENT

The consortia activities in Guinea, Australia, and Brazil offer attractive opportunities for the individual aluminum MNCs to economize on investment, as well as improve safety. The use of larger consortia has increased in recent years, reducing size of individual investments and risk. The enormous size of the bauxite reserves in the three countries permits the consortia larger-scale, economical operations. In Brazil, the government is putting in quite large investment funds itself, thus adding to security and also reducing individual MNC investment. It has just been announced that Brazil and Japan will jointly invest in a 51–49 percent arrangement for a $1.4 billion complex of bauxite-alumina-aluminum in the Amazon jungle. In Australia, also, local funds are being invested with the same effects. In all the countries, the international banking community and institutions—World Bank-IFC, Export-Import Bank, European finances, etc.—have been brought in. They provide funds and their presence gives assurances against improper or arbitrary take-overs. They provide a "good housekeeping" seal of approval for MNC investors. Insurance from Overseas Private Investment Corporation (OPIC) is available for Guinea and Brazil bauxite investments, and is not for Jamaica at present. The aggregate investment over long-term purchase contract arrangements in Guinea, Brazil and Australia appear to offer more attractive investment conditions, as well as long-term supply security, than Jamaica following its recent take-over and levy orders.

OPTIONS

There is some evidence that the companies are increasing their options, beyond the above diversification. In each of the initial arrangements in

Guinea, Brazil, and Australia, further stages of expansion are contemplated. Very large, economical expansions in future capacities are possible in these countries. In numerous other countries, the companies are engaged in less intensive exploration and negotiation. We have found press or journal reports of possible future developments in France, Greece, Yugoslavia, Cameroon, Ghana, Guinea, Malagasy, Malawi, Mali, Sierra Leone, India, Indonesia, the Philippines, Turkey, Costa Rica, Dominican Republic, Haiti, the United States, Brazil, French Guiana, Guyana, Australia, British Solomon Islands, and Fiji.

NON-BAUXITIC ORES

The world volume of high-quality bauxite reserves is very large—hundreds of years' supply at present rates of use. The world volume of aluminum in other ores and clays is a hundred times greater. America is lavishly endowed with these ores. A single leasehold of anorthosite which Alcoa recently acquired in Wyoming is said to contain enough aluminum to satisfy world demand for more than fifty years. Techniques for extracting aluminum from alternative ores are well known. The Soviet Union produces some of its aluminum from one of these ores.

The cartel problem is that such ores are another limitation on cartel ability to raise bauxite prices without losing market.

Until the recent doubling or trebling of bauxite prices, bauxite had a clear advantage in cost to produce alumina. This is why the MNCs have preferred bauxite over other ores. Now, at present bauxite prices, the other aluminum ores are becoming competitive in cost. All of the companies are now developing knowledge on the economics of refining alumina from alternative ores. In addition, the U.S. Bureau of Mines is testing out six processes in a joint research venture, partly financed by the companies. Several of the company efforts are at sizes just below small commercial scale. A full-size plant to produce 400,000 tons of alumina a year plus 500,000 tons of fertilizer from alunite is planned to be put into construction this year or next.

The London-based Commodities Research Unit (CRU) has estimated, in a November 1974 document for OPIC, that there will be several U.S. alumina plants in operation by 1980, based on alternative ores, and that thereafter domestic non-bauxite capacity will increase rapidly. CRU also stated that Pechiney will probably switch to alternative ores for all new capacities in France by 1980 if their H+ process test plants (joint with Alcan) are successful.

My own observation is that firms are reluctant to make definite decisions at this time to go to full-size plants for utilization of ores alternative to bauxite. I hesitate to differ with CRU, which has close communication with and deep understanding of the aluminium industry. But so it appears to me.

While the techniques for producing from other ores are known, the economics of commercial production in large-scale plants are not. To learn this it is necessary to build very small-scale plants, then larger ones, then full-size commercial plants. It is a big commitment.

Continued reliance on bauxite offers companies three big advantages. Alumina from bauxite still seems to be cheaper even though much higher than before 1974. Over a decade or more, as experience is gained with non-bauxitic ores on the "learning curve," cost to produce alumina might fall below cost from bauxite. But for the first "generation" of ore and alumina facilities, the cost is likely to be moderately higher. Provided it does not get caught in a bauxite price squeeze, the longer a firm can wait, the better, particularly if it can learn from the development efforts of Bureau of Mines, Alcoa, Pechiney, Alcan, and National Southwire-Earth Sciences. Second, the new process alumina plants and non-bauxite facilities would require considerably larger investment funds than continued reliance on bauxite, and capital funds are short. Third, the cartel and its members may weaken, and the future prices and tax levies of bauxite may not only not increase but may fall from present levels.

With respect to non-bauxite sources, my observation is that the companies in general are trying to "stay loose" and flexible. There are variations among them, reflecting their bauxite positions, geography, capital availability, technological character, industry position, and expectations. Some are developing mini-commercial plants, others not. Some are acquiring properties. Except for Earth Sciences–National Southwire, none has yet made a major plant investment decision. And in that case, the interest is as much in fertilizer products from the joint process as in alumina; and actual construction has not started.

U.S. OIL AND COAL DEPOSITS

A research associate (Benjamin Woodbury) and I have done a preliminary, brief investigation related to U.S. non-bauxitic ores in the Western states. We identified some of these ores as located near cheap Western coal. We conceived the hypothesis that, for new supplies of aluminum, an energy cost advantage of adjacent cheap coal for alumina refining and aluminum smelting might overcome a cost disadvantage of non-bauxitic ores. This possible energy-ore location advantage has been strengthened by energy changes (great increase in oil prices, the disappearance of incremental cheap hydroelectric power in the Pacific Northwest and TVA areas, the disappearance of cheap gas in the Gulf Coast area, and large price increases in environmentally acceptable Eastern coal), by bauxite price increases, and by increased risks in foreign investments in LDCs.

If or when U.S. firms do move to utilization of non-bauxitic U.S. ores, it seems to us that location of some of the alumina and aluminum facilities at places of favorable coal, ore, and transportation will become economically advantageous. If this occurred, it would be a major relative shift of aluminum-smelting facilities, which until recently have been located in the Pacific Northwest, Gulf Coast, and Ohio Valley. They were oriented to cheap power and convenience for receipt of Australian and Caribbean bauxite or alumina. Optimal locations for incremental facilities in the aluminum industry are changing.

SUMMARY

Without formal cartel action, but with cartel communication and spirit, the LDC members have strongly followed Jamaica leadership in revising bauxite taxes upward and in other new economic order measures. The result has been to double delivered bauxite prices and double to treble f.o.b. prices. The levies and other measures now return to the governments amounts equal to 100 percent to 200 percent of the costs of production. Further, the tax is "indexed" (proportioned) to a DC industrial goods price (aluminum ingot), and in principle provides some inflation protection. In effect, a paramenter or convention in the bauxite market place has been changed. Formerly, the conventional tax was $1 or $2 a ton of bauxite. Now it has becomes $6 to $12, except in Australia. Overall, this would appear to be a significant cartel group success.

However, there are clouds on the cartel horizon. The absence of levy in Australia has moved Australia from competitive disadvantage because of transport distance to advantage in supply of alumina. Guinea also has acquired competitive advantage by imposing a smaller levy. Both Australia and Guinea are increasing sales and relative share of the market, while the Caribbean members have, approximately, constant sales and declining market share. Further, very large new reserves are being developed for supply in Brazil, which is not a member of the cartel and whose objective is large expansions of minerals capacities and supplies. And in the wings is Cameroon, a potential Guinea-type competitor. In these circumstances Caribbean desires for further increases in levies and expansions of output are likely to be frustrated, and markets for some of the countries may even decline. Under these pressures, levies may erode, if the countries try to hold or improve sales.

In summary, levies are likely to stay where they are. If they do not, it is more probable that they will decline than increase. Cartel or oligopoly price discipline seems very weak. The cartel may have already had its great days.

CONCLUSION

The essential question is as follows:
(a) Assume the present cartel-led levies and prices;
(b) assume growth in demand;
(c) assume fringe suppliers and some cartel members who are increasing output and, if necessary, will shade price significantly in order to sell; and
(d) assume cartel leaders who are trying to maintain the price, and are willing to reduce output to do so.

Will the cartel leaders reduce output sufficiently (d) so that the fringe and other member suppliers can clear their outputs without shading price significantly (c)?

- If the answer is yes, the price and the cartel will hold firm.
- If the answer is no, then competitive price declines will occur; and, if not rectified with a more stable price in a reasonable time, cartel power and influence will dissolve.

In the petroleum case, it has been true for several years and appears likely for a few more that Saudi Arabia, Kuwait and others are willing to reduce output sufficiently (d) to permit other oil supply and competitive fuels to sell out, at present cartel prices or only moderately lower. In fact, the cartel real price of oil has been eroded by a third by inflation. But the cartel is holding at the money price.

In the bauxite case, already other suppliers—Guinea, Australia, Brazil, Guyana, and Indonesia—have undercut Jamaica. And already Jamaica has apparently reduced its levy in an effort to hold volume. The prospects are for more supply increases than the leaders can absorb in output reduction. There is too much bauxite reserve, at too many places, producible at too low costs, in too many nations which want revenues, to contain supply at Jamaica's announced tax levies. It is beginning to appear that there may be too much bauxite available even to contain supply at moderately lower levies. And looming in the background are the mammoth resources of non-bauxitic aluminous ores, some of them located near lowcost energy resources.

In conclusion, the bauxite-producing countries are better off than they were formerly in that improved tax levies have now become somewhat conventional. The cartel itself, however, seems to be relatively ineffective and possibly crumbling.

ACKNOWLEDGMENT

This paper is dated 1976. It is primarily drawn from a larger research effort begun during the author's tenure as Fellow, Woodrow Wilson International Center for Scholars, Smithsonian Institution, Washington, D. C. The author is grateful to the Center for the Fellowship research opportunity and warm associations. He is grateful to the International Institute for Applied Systems Analysis, Vienna, also, for congenial fellowship and for his research activity on energy and resources availabilities. The author wishes to point out that his bauxite cartel research is still incomplete, and that neither organization is responsible for views or errors herein.

T.F. Yen
EDITOR

Energy Sources

An International Interdisciplinary Journal of Science and Technology

Now in its third year of publication, *Energy Sources* has become a respected source of advanced information and sophisticated thinking about the wide range of problems and alternatives that comprise the planet's energy crisis.

The journal bridges gaps among scientists, engineers, and technologists involved in the development of new sources of energy into useful and efficient fuels. *Energy Sources* is a forum for members of several disciplines working toward a common goal: workable solutions to a worldwide, worsening, shortage of energy.

A sampling of recent articles includes:

- Prospects for Coal as a Direct Fuel and Its Potential Through Application of Liquefaction and Gasification Technology *(J.R. Bowen)*
- Oxidation of Lignite into Water-Soluble Organic Acids *(D.K. Young and T.F. Yen)*
- The Influence of the Properties of Coal on Their Conversion Into Clean Fuels *(A. Davis, W. Spackman, and P.H. Given)*
- Wind Power Feasibility *(J.P. LeBoff)*
- The Fischer Assay: Standard for the Oil Shale Industry *(R.N. Heistand)*
- Capital and Electric Production Costs for Geothermal Power Plants *(H.E. Klei and F. Maslan)*
- A Critique of the Nationwide 55 mph Speed Limit *(D. Rapp)*
- Hazardous and Toxic Air Pollutants *(M. Samfield)*

Energy Sources is published quarterly. Each volume contains approximately 400 pages. The subscription price for Volume 3 is $36.00. The subscription price for Volume 4 (beginning Fall, 1978) is $44.00. Personal subscriptions (payable by personal check only) are available at a 50% discount.

For subscriptions outside North America, please add $4.00 per volume for postage and handling.

Crane, Russak & Company, Inc.
347 Madison Avenue • New York 10017

OTHER ANNUAL SERIES OF INTEREST FROM JAI PRESS INC.

Consulting Editor for Economics: Paul Uselding, University of Illinois

ADVANCES IN APPLIED MICRO-ECONOMICS
Series Editor: V. Kerry Smith, Resources for the Future, Washington, D.C.

ADVANCES IN ECONOMETRICS
Series Editors: R. L. Basmann, Texas A & M University, and George F. Rhodes, Colorado State University

ADVANCES IN ECONOMIC THEORY
Series Editor: David Levhari, The Hebrew University

ADVANCES IN THE ECONOMICS OF ENERGY AND RESOURCES
Series Editor: Robert S. Pindyck, Sloan School of Management, Massachusetts Institute of Technology

APPLICATIONS OF MANAGEMENT SCIENCE
Series Editor: Randall L. Schultz, Krannert Graduate School of Management, Purdue University

RESEARCH IN AGRICULTURAL ECONOMICS
Series Editor: Earl O. Heady, Director, The Center for Agricultural and Rural Development, Iowa State University

RESEARCH IN CORPORATE SOCIAL PERFORMANCE AND POLICY
Series Editor: Lee E. Preston, School of Management and Center for Policy Studies, State University of New York, Buffalo

RESEARCH IN ECONOMIC ANTHROPOLOGY
Series Editor: George Dalton, Northwestern University

RESEARCH IN ECONOMIC HISTORY
Series Editor: Paul Uselding, University of Illinois

RESEARCH IN EXPERIMENTAL ECONOMICS
Series Editor: Vernon L. Smith, College of Business and Public Administration, University of Arizona

RESEARCH IN FINANCE
Series Editor: Haim Levy, School of Business, The Hebrew University

RESEARCH IN HEALTH ECONOMICS
Series Editor: Richard M. Scheffler, University of North Carolina, Chapel Hill and the Institute of Medicine, National Academy of Sciences

RESEARCH IN HUMAN CAPITAL AND DEVELOPMENT
Series Editor: Ismail Sirageldin, The Johns Hopkins University

RESEARCH IN INTERNATIONAL BUSINESS AND FINANCE
Series Editor: Robert G. Hawkins, Graduate School of Business Administration, New York University

RESEARCH IN LABOR ECONOMICS
Series Editor: Ronald G. Ehrenberg, School of Industrial and Labor Relations, Cornell University

RESEARCH IN LAW AND ECONOMICS
Series Editor: Richard O. Zerbe, Jr., SMT Program, University of Washington

RESEARCH IN MARKETING
Series Editor: Jagdish N. Sheth, University of Illinois

RESEARCH IN ORGANIZATIONAL BEHAVIOR
Series Editors: Barry M. Staw, Graduate School of Management, Northwestern University, and Larry L. Cummings, Graduate School of Business, University of Wisconsin

RESEARCH IN PHILOSOPHY AND TECHNOLOGY
Series Editor: Paul T. Durbin, Center for the Culture of Biomedicine and Science, University of Delaware

RESEARCH IN POLITICAL ECONOMY
Series Editor: Paul Zarembka, State University of New York, Buffalo

RESEARCH IN POPULATION ECONOMICS
Series Editors: Julian L. Simon, University of Illinois, and Julie DaVanzo, The Rand Corporation

RESEARCH IN PUBLIC POLICY AND MANAGEMENT
Series Editors: Colin C. Blaydon, Institute of Policy Studies and Public Affairs, Duke University, and Steven Gilford, Chicago

ALL VOLUMES IN THESE ANNUAL SERIES ARE AVAILABLE AT INSTITUTIONAL AND INDIVIDUAL SUBSCRIPTION RATES. PLEASE WRITE FOR DETAILED BROCHURES ON EACH SERIES

A 10 percent discount will be granted on all institutional standing orders placed directly with the publisher. Standing orders will be filled automatically upon publication and will continue until cancelled. Please indicate which volume Standing Order is to begin with.

JAI PRESS INC.
P.O. Box 1285
165 West Putnam Avenue
Greenwich, Connecticut 06830

(203) 661-7602 Cable Address: JAIPUBL.

Energy Systems *and* Policy

an international interdisciplinary journal

S. William Gouse, Jr.

EDITOR-IN-CHIEF

This new journal provides an international forum for examining large-scale energy system problems. It satisfies a long-felt need for a vehicle to present and discuss the complex multidisciplinary interactions that take place between physical science technology, economic, social and political sciences, international relations, aspirations of developing nations, and the inter-relationship between these disciplines and existing energy policy as well as what might be in the future. All aspects of energy systems are considered, from source to ultimate consumption.

Selected articles in early issues of ENERGY SYSTEMS AND POLICY included: *A Sulfur Emission Tax and the Electric Utility Industry; The Potential of Coal to Meet the Energy Crisis; A Program for Maximizing U.S. Energy Self-Sufficiency; Energy Rationing and Energy Conservation: Foundations for a Social Policy; Scenarios of Electric Utility Fuel Use in the United States, 1980 and 1990; Conservation Options for Air Transport; Discussion of Uranium Availability and the Breeder Decision; Elasticity of Demand for Gasoline;* and *Environmental Policies for Electricity Generation.*

ENERGY SYSTEMS AND POLICY is published quarterly. Each volume contains approximately 400 pages. The subscription price for Volume 2 is $36.00. The subscription price for Volume 3 (beginning Fall, 1978) is $44.00. Personal subscriptions (payable by personal check only) are available at a 50% discount.

For subscriptions outside North America, please add $4.00 per volume for postage and handling.

Crane, Russak & Company, Inc.
347 Madison Avenue, New York, N.Y. 10017

Advances in the Economics of Energy and Resources

A Research Annual

Series Editor: **Robert S. Pindyck, Sloan School of Management Massachusetts Institute of Technology.**

Advances in the Economics of Energy and Resources brings together new research results covering a wide spectrum of problems in energy and resource economics and policy. Each volume will be structured around core problems in this field. **Volume 1: The Structure of Energy Markets,** covers aspects of energy markets ranging from the behavior of individual firms, to the behavior of the entire energy sector, to the interactions between the energy sector and the macroeconomy. The coverage ranges from domestic (U.S.) energy markets and energy policy to international problems in energy. **Volume 2: The Production and Pricing of Energy Resources,** covers a wide range of problems in energy resource supply, pricing and management. Topics treated include the statistical forecasting of resource discovery rates, the optimal exploitation and pricing of resources, the role of supply in world resource markets and the relationship between resource markets and the macroeconomy. The papers included in these and future vllumes represent new and significant research results in the economics of energy and resources.

Volume 1: The Structure of Energy Markets

April 1979 Cloth 325 pages Institutions: $ 27.50

ISBN 0-89232-078-8 Individuals: $ 14.00

CONTENTS: Energy-Economy Interactions: The Fable of the Elephant and the Rabbit? *William W. Hogan and Alan S. Manne, Stanford University.* **Input Prices: Substitution and Product Inflation,** *John R. Moroney, Tulane University and Alden Toevs, Louisiana State University.* **Substitution Among Energy, Capital and Labor Inputs in U.S. Manufacturing,** *Robert Halvorsen and Jay P. Ford, University of Washington.* **The Stanford Energy/Economic Model,** *Thomas J. Connally, George B. Dantzig and Shailendra C. Parikh, Stanford University.* **New Car Efficiency Standards and the Demand for Gasoline,** *James L. Sweeney, Stanford University.* **Factors Leading to Structural Change in the U.S. Oil Refining Industry in the Postwar Period,** *Stephen C. Peck, Electric Power Research Institute, Palo Alto, and Scott Harvey, University of California—Berkeley.* **Integration and Innovation in the Energy Markets,** *David J. Teece, Stanford University.* **Economics of the Throwaway Nuclear Fuel Cycle,** *Martin Baughman and John Gordon, University of Texas—Austin.* **Prospects for Nuclear Power in the Developing Countries,** *Alan M. Strout, Massachusetts Institute of Technology.* **Financial Markets and the Adjustment to Higher Oil Prices,** *Tamir Agmon, Tel Aviv University, James L. Paddock and Donald R. Lessard, Massachusetts Institute of Technology.*

A 10 percent discount will be granted on all institutional standing orders placed directly with the publisher. Standing orders will be filled automatically upon publication and will continue until cancelled. Please indicate which volume Standing Order is to begin with.

JAI PRESS INC., P.O. Box 1285, 165 West Putnam Avenue, Greenwich, Connecticut 06830.

Telephone: 203-661-7602 Cable Address: JAIPUBL

RESEARCH IN PHILOSOPHY AND TECHNOLOGY

A Research Annual

Series Editor: Paul T. Durbin, Philosophy Department and Center for the Culture of Biomedicine and Science, University of Delaware.

Bibliography Editor: Carl Mitcham, St. Catharine College, St. Catharine, Kentucky.

OFFICIAL PUBLICATION OF THE PHILOSOPHY AND TECHNOLOGY GROUP

The series will serve three functions in the advancement of the philosophy of technology movement in American academic circles. Through its annual bibliographical update under the direction of Carl Mitcham, it will become the place to look for bibliographical leads, documentation, and all the other things a good bibliography serves for. It will serve as the outlet for the proceedings of a series of philosophy of technology conferences and symposia, with a major conference at least every other year. It will serve as a sort of substitute for a journal in the field, with submitted papers not intended for one of the conferences or symposia welcome at any time.

Volume 1. Published 1978 Cloth 400 Pages Institutions: $28.50
ISBN NUMBER 0-89232-022-2 Individuals: $15.00

CONTENTS: Introduction to the Series. Introduction to Volume One.

Part 1—Method, Descriptive Framework, and a Practical Program for Philosophy of Technology: Culture and Technology, Joseph Margolis, Philosophy Department, Temple University. **The Problem of Scale in Human Life: A Framework for Analysis,** Robert E. McGinn, Values, Technology, and Society Program, Stanford University. **Technology, Mass Movements, and Rapid Social Change: A Program for the Future of Philosophy of Technology,** Joseph Agassi, Philosophy Department, Boston University and Tel Aviv University.

Part II—The University of Delaware Conference, 1975: Toward a Social Philosophy of Technology, Paul T. Durbin, Philosophy Department, Center for the Culture of Biomedicine and Science, University of Delaware. **The Explanation of Technology,** Albert Borgmann, Philosophy Department, University of Montana. **Values in Science,** Willis H. Truitt, Philosophy Department, University of South Florida. **Technology as Ideology,** Kai Nielsen, Philosophy Department, Brooklyn College and University of Calgary. **Humanization of Technology: Slogan or Ethical Imperative?** Edmund Byrne, Philosophy Department, Indiana University and Purdue University at Indianapolis. **What is Technology?** Robert E. McGinn, Values, Technology and Society Program, Stanford University. **Shifting from Physical to Social Technology,** Joseph Agassi, Philosophy Department, Boston University and Tel Aviv University. **The Cognitive Dimension of Technological Change,** Stanley R. Carpenter, Department of Social Sciences, Georgia Institute of Technology. **Types of Technology,** Carl Mitcham, St. Catharine College, St. Catharine, Ky.

Part III—Bibliographical Update: Philosophy of Technology, 1972–1974, Carl Mitcham.

Index.

Research in Philosophy and Technology

Volume 2. September 1979 Cloth Ca. 425 pages Institutions: $ 32.50
ISBN 0-89232-101-6 Individuals: $ 16.50

CONTENTS: Introduction to Volume Two: The International Cast of the Philosophy and echnology Movement.

Part I: EUROPEAN CONTRIBUTIONS. **Euthyphronics and the Problem of Adapting Technical Progress to Man,** *Jozef Banka, Katowice, Poland.* **Toward an Interdisciplinary and Pragmatic Philosophy of Technology,** *Hans Lenk and Gunter Ropohl, University of Karlsruhe.* **Fear and nsight in French Philosophy of Technology,** *Daniel Cerezuelle, Universite de Grenoble.*

Part II: CONFERENCE PAPERS, **1976-1977. Freedom and Determinism in a Technological Setting,** *Albert Borgmann, University of Montana.* **The Normative Side of Technology,** *Edmund Byrne, Indiana University and Purdue University.* **Art in a Technological Society,** *Phillip Frandozzi, University of Montana.* **Marx, Machinery and Alienation,** *Bernard Gendron, University of Wisconsin Milwaukee and Nancy Holmstrom, University of Wisconsin - Madison.* **Technological Culture and he End of Philosophy,** *Michael Zimmerman, Newcomb College, Tulane University (Comment by Robert McGinn, Bell Laboratories: Reply by Zimmerman).*

Part III: HISTORICO-PHILOSOPHICAL STUDIES. **The Concepts of 'Nature' and 'Technique' ccording to the Greeks,** *Wolfgang Schadewaldt, Universitat Tubingen.* **On the Nature of Nature,** *acob Klein, St. John's College, Annapolis.* **Documentation: Early Analyses of Machines in the rench Intellectual Tradition,** *Alfred Espinas, Jacques Lafitte and Simone Weil.* **On the ntagonism Between Philosophy and Technology in Germany and Austria,** *Werner Koenne, ustralian Elektrizitat Wirtschafts-Aktiengesellschaft, Vienna.* **Heidegger and Marcuse: Technology s Ideology,** *Michael Zimmerman, Newcomb College, Tulane University.*

art IV: REVIEW AND BIBLIOGRAPHY. **Philosophy and Technology and the Verein Deutscher ngenieure,** *Alois Hunting, VDI Wulfrath, West Germany.* **Philosophy of Technology in France: Overview and Bibliography, 1972-1977,** *Jean-Claude Beaune, Universite de Clermont.* **Hendrik an Riessen and Dutch Neo-Calvinist Philosophy of Technology, Hendrik van Riessen: Biography and Selected Bibliography,** *Donald Morton, Free University, Amsterdam.* **The Structure of Technology.** *Hendrik van Riessen, Free University, Amsterdam.* **Technology and ulture,** *Hendrik van Riessen, Free University, Amsterdam.* **Toward a Neo-Calvinist Philosophy of echnology: Review of E. Schuurman, Technology and Deliverance,** *Donald Morton, Free niversity, Amsterdam.* **Technology Assessment: A New Type of Science?** *Frederick A. Rossini, Georgia Institute of Technology.* **Supplementary Bibliography,** *Carl Mitcham and Jim Grote, St. atherine College.* **Kenneth Sayre on Information-Theoretic Models of Mind,** *K. S. Shrader-echette, University of Louisville.*

APPENDIX. **Author Index for Current Bibliography in the Philosophy of Technology, 973-1974, Research in Philosophy and Technology, Volume 1 (1978).**

A 10 percent discount will be granted on all institutional standing orders placed directly with the publisher. Standing orders will be filled automatically upon publication and will continue until cancelled. Please indicate which volume Standing Order is to begin with.

JAI PRESS INC., 165 West Putnam Avenue, P.O. Box 1285 Greenwich, Connecticut 06830.

elephone: 203-661-7602 Cable Address: JAIPUBL